THE WATER DIARIES

Cyclones, flash floods, droughts, and pollution batter the aspirations of people living at the sharp end of water insecurity. By charting the daily water use behaviour of people in Kenya and Bangladesh for a year, this book explores the intersecting drivers of global water risks and the spatial and seasonal inequalities. Comprising a clear methodological chapter and four detailed case studies of both urban and rural areas, it critically reviews existing policy and institutional design, arguing for a new architecture in allocating risks and responsibilities fairly and effectively between government, communities, enterprises, and water users. In identifying the risks and potential responses for policy and investment action, it provides theoretical insights and a practical guide to developing more effective policy in Kenya and Bangladesh, with solutions that will be applicable in other regions facing similar challenges. This title is also available as Open Access on Cambridge Core.

SONIA HOQUE is a Bangladesh-born environmental social scientist, with eight years of postdoctoral research on water security and poverty under the UK Foreign Commonwealth Development Office (UK FCDO)-funded REACH Programme at the University of Oxford. She has worked in Bangladesh, Ethiopia, and Kenya, focusing on the inequalities in household and individual experiences of water risks related to rural and urban drinking water services and river pollution by the global fashion industry.

ROB HOPE is Professor of Water Policy at the Smith School of Enterprise and the Environment at the University of Oxford. He directed the UK FCDO REACH Programme between 2015 and 2024 that has improved water security for over 10 million vulnerable people in Africa and Asia. He co-founded Uptime in 2018 which has issued results-based contracts to guarantee reliable drinking water for over 5 million rural people in 17 countries in 2024. He has published over 100 academic papers.

THE WATER DIARIES

Living with the Global Water Crisis in Bangladesh and Kenya

SONIA HOQUE
University of Oxford

ROB HOPE
University of Oxford

Shaftesbury Road, Cambridge CB2 8EA, United Kingdom

One Liberty Plaza, 20th Floor, New York, NY 10006, USA

477 Williamstown Road, Port Melbourne, VIC 3207, Australia

314–321, 3rd Floor, Plot 3, Splendor Forum, Jasola District Centre, New Delhi – 110025, India

103 Penang Road, #05–06/07, Visioncrest Commercial, Singapore 238467

Cambridge University Press is part of Cambridge University Press & Assessment,
a department of the University of Cambridge.

We share the University's mission to contribute to society through the pursuit of
education, learning and research at the highest international levels of excellence.

www.cambridge.org
Information on this title: www.cambridge.org/9781009299589

DOI: 10.1017/9781009299596

First published 2025

A catalogue record for this publication is available from the British Library

A Cataloging-in-Publication data record for this book is available from the Library of Congress

ISBN 978-1-009-29958-9 Hardback

Contents

Figures

Table

Preface

This book is about a daily global puzzle. How do vulnerable people facing droughts, floods, or pollution make water choices each day? They are the people at the very sharp end of the climate crisis compounded by economic instability, weak governance, and public health shocks. They often feature in the news as disaster strikes as homes and livelihoods are brutally washed away, local rivers brim with disease and rubbish, or rain fails to fall for months and months on end. With limited resources and assistance, over two billion marginalised people have to make daily choices to find and use water for drinking, cooking, washing, cleaning, hygiene, or watering livestock. We believe by better understanding these daily water choices that we can design and shape better policy and interventions to help improve water security for hundreds of millions of people in Africa and Asia.

As two researchers who have conducted many households surveys, interviews, and focus group discussions over two decades in more than a dozen countries, we know the strengths and limits of existing research methods. We have been inspired by new ideas and approaches, such as the compelling book, *Portfolios of the Poor*, which charted the daily expenditure of people earning less than USD 2 per day in Bangladesh, India, and South Africa. The findings revealed the incredible sophistication of 250 families managing their limited and variable income based on fortnightly interviews. The insights provided clear and actionable evidence for better policy and practice that has been taken up widely since. It gave us a framework to see how these methods might be adapted and advanced to create a suitable water diary method.

Water diaries are not a new idea. Researchers have previously implemented the diary method to monitor water behaviour in resource-poor settings. However, limited resources constrained the scope and ambition of these earlier studies. We were fortunate to be in a long-term and well-funded research programme that could allow us to be more ambitious and document a full year of the daily water choices of hundreds of families in multiple countries.

However, it was not only intellectual curiosity that started the diary work. Our research programme faced a major headache. The REACH Programme was committed to improving water security for 10 million poor people by understanding how climate and hydrological systems interact with water needs for drinking water, irrigation, industry, and the environment. The idea of water security is precisely about these interactions between water resources and water services and the distribution of impacts for different people across space and time. Our interdisciplinary research depended on bringing together ideas, methods, and data to examine water insecurity in different geographies facing significant but uncertain threats. While our climate colleagues could roll out hourly data on temperature, rainfall, and more, the social scientists struggled to rustle up one household survey in a year. The water diaries provided a bridge across the data gap.

In 2017, we started the water diary work in Kenya. We had worked there for many years in a remote and rural location about 160 km east of Nairobi. Hot, dry, and dusty, it was a good place to pilot the method and develop our plans with good local partners. Over the next two years, we rolled out the diaries in three other locations in Bangladesh and Kenya. The choice of countries reflected our interests, expertise, and local partner networks, as well as providing a diversity of water insecurity challenges from river pollution in Dhaka to cyclonic storms in coastal Bangladesh, and the drought cycles of rural Kenya. There is no claim that these locations represent water security globally, though the geographies capture many common and growing challenges around the world.

The methods we developed may be of interest to researchers and students. As teaching staff at the University of Oxford, we share and discuss these methods as part of our teaching on research methods for our postgraduate students. We have written a methodological chapter that some readers may wish to turn to first to better understand how the data were generated. The method has generated wider interest with UNICEF applying the method in refugee camps in Ethiopia. Other groups may find the approach to be applicable to their work on tracking water collection, use, and expenditure in a simple and relatively low-cost fashion.

For the policy community, the book poses questions on current practices and future strategies. Monitoring water security is a difficult process. *The Water Diaries* reveals how daily practices are more nuanced and variable than standard data from household surveys. The progress in global monitoring of drinking water security is to be applauded, but there are limitations, particularly with the complex synergies between seasonal and sub-seasonal rainfall patterns and daily water use. As the climate crisis unfolds, this is a major policy gap to credibly and effectively direct climate finance to the most vulnerable. The slippery slope of pedestrian policy is to focus on valuing what we measure, rather than measuring what we value. *The Water Diaries* provides unique access to daily water practices that contributes to a deeper understanding of social and cultural practices that do not neatly align to policy imposed from above.

We hope this book provides ideas and insights to understand and respond more effectively to the hundreds of millions of people living through a daily water crisis across the world. The positive news is much can be done to improve people's lives. New initiatives and investments have emerged in response to the partnerships we have shared with governments and donors in the geographies where we work. We sincerely hope that more can be scaled up fairly and quickly to improve and sustain water security for everyone.

Abbreviations

BDT	Bangladeshi Taka
BIWTA	Bangladesh Inland Water Transport Authority
BUET	Bangladesh University of Engineering and Technology
DoE	Department of Environment
DPHE	Department of Public Health and Engineering
DWASA	Dhaka Water Supply and Sewerage Authority
icddr,b	International Centre for Diarrhoeal Disease Research, Bangladesh
KES	Kenyan Shillings
LGD	Local Government Division
LOWASCO	Lodwar Water and Sanitation Company
MDG	Millennium Development Goal
MICS	Multiple Indicator Cluster Survey
NGO	non-governmental organisation
SDG	Sustainable Development Goal
UNICEF	United Nations Children's Fund
USD	United States Dollar
WASREB	Water Services Regulatory Board
WHO	World Health Organization

1
Introduction

1.1 Global Water Risks and Local Practices

One in four people lack safe and reliable drinking water, most of whom live in Asia and Africa (WHO/UNICEF, 2023). Elevated levels of arsenic and fluoride are slowly and silently poisoning tens of millions of groundwater dependent populations in the Indo-Gangetic delta of South Asia (Fendorf et al., 2010) and the East African Rift Valley (Ligate et al., 2021, Reimann et al., 2003), respectively. Rising global temperatures are invigorating cyclones and storm surges in the Indian Ocean. In May 2020, cyclone Amphan inundated coastal settlements in India and Bangladesh, exacerbating the hardships of the COVID-19 lockdown (Kumar et al., 2021). In the Horn of Africa, multiple years of failed rainfall have resulted in the worst drought in 70 years in 2022, causing water and food scarcity for 25 million people and tens of millions of livestock (OCHA, 2023). The alarming state of water risks across the world are increasingly portrayed by the media. Photos of emaciated children standing by animal carcasses, men boating along indigo tainted waters, or women wading through waist-deep flood waters narrate the diverse water risks experienced every day.

Stories from North America and Europe are increasingly making headlines as well. In 2014, a major public health crisis unfolded in the city of Flint, Michigan, after a switch in the municipal water source resulted in insufficient corrosion control in aging pipes, leading to high levels of lead in the water supply (Pauli, 2019). In the UK, since 2020, the issue of river pollution has gained increased political attention as private water companies have been found to be regularly releasing untreated sewage into rivers (House of Commons, 2022).

Over two billion people live with security risks every day. The wealthy can often buy their way out of water security risks, the poor have fewer options. The risks the poor face each day can be unpredictable or relentless with no quick or simple solution. We set out to document the risks and responses that poor people face through water diaries in Bangladesh and Kenya, two countries with different but extensive water security challenges. We use the term 'poor' as shorthand to reflect individual, household and community vulnerability, exclusion, and deprivation. Equally, their strategies, practices and creativity reflect the resilience, ingenuity, and stoicism of the poor in a world of increasing water security risks which they did not engineer, but for which they face the greatest risks to their lives and livelihoods.

Water risks are generally defined by the privileged and experienced by the poor. Attribution of and blame for water risks can take many forms with little account for the local lived practices which attempt to adapt and mitigate the most harmful outcomes. In the case of water, these outcomes can relate to human health and well-being, ecosystems, or economic development, and can be driven by multiple environmental, financial, institutional, and social factors operating at different spatial and temporal scales (Hoque et al., 2019). Environmental factors encompass the geographical and seasonal distribution of water resources and hazards, as exemplified by the public health risks from naturally occurring arsenic or fluoride in groundwater, and crop failures and livestock deaths from delayed onset or failed rainy seasons. Financial drivers of water risks range from capital expenditure gaps in water supply or wastewater treatment infrastructure to poor recovery of operational costs due to non-payment of user fees. Institutional arrangements define how responsibilities for risk mitigation are allocated among national and local governments, private sector, donor organisations and households, ranging from day-to-day service provision to water sector regulation and monitoring (Hope and Rouse, 2013). Social factors, such as poverty, gender, race and power dynamics, can disproportionately put certain population groups or individuals more at risk than others.

Global policies to address water risks have undergone several shifts in ideologies in the past few decades, depending on how the drivers of risks are framed (Hope et al., 2019, Gunawansa et al., 2013). Investments in large-scale infrastructure such as dams and centralised piped schemes gained momentum from the mid twentieth century to bring water to the people in growing cities, while deltaic floodplains were engineered with embankments, sluices, and canals to protect coastal populations from floods and storm surges. Water was viewed as a social good, with responsibilities for financing and management resting with the public sector. For nations emerging from colonial domination, overseas development aid served as the main funding source, allowing international donor organisations to intervene in national development policies based on western ideologies and practices. Infrastructure-led solutions also permeated the rural water sector in low-income countries, with low-cost handpump technologies being popularised to shift rural populations from surface water to groundwater sources. The type of water technology used became synonymous with water safety, with access to improved sources being the global policy target until today.

Since the late 1980s, the focus of risk mitigation expanded from a sole emphasis on financial factors to include institutional arrangements for allocating responsibilities and blame. Alongside the UK's privatisation of the water sector, various forms of public–private partnerships were implemented in large cities in Latin America, the Middle East, and the Asia-Pacific (Bakker, 2011). Neoliberal policies promoted water as an economic good, with the focus being on getting the prices right to ensure cost recovery and affordability. Decentralisation drives responsibilities for rural water provision to local governments, often lacking financial and institutional capacity to fulfil their mandates. Community-based management became the standard model for operation and maintenance of rural water infrastructure based on demand-driven ideologies. Unlike urban areas with subsidised piped services managed by utilities, rural users are left to their own devices, having to gather money and spare parts for repairing their pumps and pipes. An extensive body of evidence from Asia and Africa shows that the community-managed model has yielded unsatisfactory outcomes, as waterpoints are poorly managed and often abandoned after a few years, with the expected lifetime rarely achieved (Foster, 2013, Whittington et al., 2009, Harvey and Reed, 2007). Uncoordinated infrastructure investments by governments and donors create a complex tapestry of water sources often located adjacent to existing waterpoints in a graveyard of well-meaning intentions.

Global and national monitoring of progress in water services in the twenty-first century have been defined by the United Nations' Millennium Development Goals (MDG) and Sustainable Development Goals (SDG). In line with the infrastructure-led provision approach, MDG Target 7c narrowly focused on measuring 'access to an improved source'. SDG Target 6.1 expanded global ambition from access to a service delivery approach, with a focus on providing safely managed services calibrated by water quality, reliability, affordability, proximity, and equity. Yet, given the costs and logistics of large-scale data collection, nationally representative surveys and censuses still focus on the 'main source of drinking water', with services defined as 'safely managed' when a source is available on demand, free of faecal contamination and on-premises, and 'basic' when it is within 30 minutes of dwelling (WHO/UNICEF, 2017). The SDGs also include ambient water quality, targeting to reduce disposal of untreated wastes into waterbodies (SDG 6.3). However, monitoring is still at a nascent stage owing to data gaps, political leadership, and an effective monitoring system (UNEP, 2021).

While commonly quoted aggregate statistics – 2.2 billion people lacking safely managed drinking water in 2022 (WHO/UNICEF, 2023) or women in Africa spend 40 billion hours a year in collecting water (UNDP, 2006) – paint the scale of global water risks, local realities are more nuanced. Access to an improved source fails to guarantee water security, as seen in the cases of arsenic, fluoride, and lead exposure despite using handpumps or piped water. Rural populations in developing countries are known to use different sources for drinking, cooking, washing, and livestock, delicately balancing the seasonal variations in water availability, water quality, costs, and distance (Elliott et al., 2019). For those surviving on limited and uncertain incomes, immediate concerns of feeding the family or paying children's school fees may need to be balanced against costly one-off investments in water supply or treatment technologies that can potentially avoid health risks in the long-term (Ray and Smith, 2021). In overcrowded urban slums, women may choose sources of lower quality or higher price to avoid queuing at shared public taps and manage time for paid work, childcare cooking, or other competing needs (Price et al., 2019). Marginalised communities living in polluted and flood-prone riverbanks may choose not to relocate to safer grounds, as doing so may mean losing proximity to markets and income opportunities (Korzenevica et al., 2024).

These examples illustrate that risks are socially constructed, and the 'tolerable' level of risks varies across societies and individuals (Grey et al., 2013) with greater need to include the 'equity' of water risks hidden in what may be tolerable for the majority. The tolerability of risks has influenced decision-making from global to local scales, since before the concept of risk was associated with 'water security'. Water quality guidelines established by the World Health Organisation (WHO), for example, prescribe 10 µg/l as the acceptable threshold for arsenic in drinking water, whereas in Bangladesh, the national standard is set at 50 µg/l as the costs to meet a lower threshold would exceed the public health benefits in a context with extremely high levels of groundwater arsenic. Household and individual water source choices are likewise driven by careful evaluation of the monetary and non-monetary costs and benefits associated with accessing sources with varied quality, distance, costs, and reliability. The 'acceptable' level of risks and water practices are shaped by people's past experiences, physical and psychological capabilities, sociocultural norms, and environmental context.

Pierre Bourdieu introduced the sociological concept of 'habitus' to reflect on people's 'practices' within the 'field' in which they operate. Habitus is 'an active presence of past experiences' (Bourdieu, 1990, p. 54), which governs the continuity and regularity of social practices by providing relatively autonomy from immediate, external constraints. The concept of habitus views one's personal experiences, sociocultural and environmental contexts as salient drivers of present practices, which tend to perpetuate into the future, reactivating in similar structures. Habitus rejects ideas of rational choice, which considers that individual decisions are guided by balanced consideration of costs and benefits of alternative choices, with the option with the highest satisfaction (utility) being chosen. These differences in individual values and preferences may not always conform to policy prescriptions of installing water supply infrastructure that users will regularly pay for and manage for their common good. Knowledge of bacterial or chemical contamination may not deter individuals from using unsafe sources or bathing in polluted rivers unless alternatives can be conveniently incorporated into people's lifestyles. Cultural norms often define whose voices and needs are prioritised in household water decisions, and how responsibilities for collecting and paying for water or managing water infrastructure are allocated between men, women, and children.

This book is about the daily water practices of individuals and households navigating various water risks – from unsafe or scarce drinking water to polluted waterbodies and extreme events – across different environmental, institutional and infrastructure settings. By charting these daily practices, we aim to better understand people's choices and constraints with a granular level of detail that allows us to rethink current policy and practice. Our work provides an empirical basis to accelerate action to reduce water risks and achieve water security in some of the most challenging geographies on the planet.

1.2 Water Diaries as a Lens to Individual Practices

The 'everydayness' of how water risks are experienced by men, women and children, whether in remote villages, in small towns or in bustling megacities, whether in the humid tropical floodplains or in the arid Sub-Saharan landscape, is what inspired us to study the water crisis through 'water diaries'. Diaries are inherently records of the mundane day-to-day activities, the details of which are likely to be erased from our memories after a short time. Fetching water from wells and pipes, or washing and bathing at the river are emblematic of individuals' deliberate and subconscious choices shaped by their *habitus*. Our water diaries were designed to capture these behaviours or *practices* related to water, operating within dynamic *fields* that also include practices by governments, donor agencies, markets, and other individuals.

We focus on four diverse study sites from two countries – Bangladesh and Kenya. The scenes of water crisis in these two countries could not be any different. Located on the low-lying floodplains of three mighty rivers – the Ganges, Brahmaputra, and Meghna – that drain the Himalayan waters into the Bay of Bengal, Bangladesh has plentiful water. Intricately woven by rivers and nourished by four months of monsoon, the country is one of the world's most densely populated places, with a population of 165 million living across an area of 147,500 km^2 (BBS, 2023). On the other hand, about 80 per cent of Kenya's land area is categorised as arid or semi-arid, characterised by two rainy seasons with low and unreliable rains feeding the seasonal rivers. With four times the land area of Bangladesh

(581,000 km^2) and only a third of its population (48 million) (KNBS, 2019a), most of Kenya's rural areas and small towns are sparsely populated. While riverine floods, cyclones and storm surges are common in Bangladesh, Kenya is affected by prolonged droughts. Despite the stark contrast in environmental risks, the two countries share some common institutional and infrastructure risks, including capital financing gaps, operation and maintenance challenges, and regulation of water resources and services.

The shared institutional and infrastructure contexts may be traced to the similar economic and political histories of the two countries, as shaped by colonisation, dependence on foreign aid post-independence, and neoliberal economic policies imposed by donor organisations. From the nomadic pastoral lifestyles of Kenyan tribes to the ebbs and flows of Bengal rivers – the 'uncontrollability' of the people and nature in these countries was a sharp contrast to colonial ideologies of modernisation. Both countries inherited colonial policy legacies of territorialisation of people by religion, ethnicity or livelihood, as well as the control of water through structural solutions. The reliance on overseas development aid post-independence allowed international financial institutions to shape national policies. Market oriented reforms were implemented in both countries to reduce public spending and mobilise private sector. The rural water sector saw a push towards decentralisation of service delivery, with the focus on scaling low-cost technologies that could be managed by users. With the handpump revolution and community-based management unfolding in the 1980s and 1990s, the infrastructure and institutional ideologies of rural water sector were redefined, a legacy which still dominates.

To study how these diverse environmental, infrastructure and institutional landscape drive daily water practices, we selected four study sites in Bangladesh and Kenya, representing the different aspects of global water crisis (Figure 1.1). These are – (a) Dhaka city, the densely populated capital of Bangladesh, where low-income riverbank settlements risk daily exposure to chemical and pathogen pollution caused by multidecadal discharge of untreated industrial and municipal wastewater into the city's rivers; (b) Khulna district, in the coastal floodplain of southwestern Bangladesh, where rural communities in embanked islands suffer from groundwater salinity and episodic shocks from cyclones and storm surges; (c) Kitui county, representing the sparsely populated semi-arid rural landscapes of Kenya, where low rainfall subject to inter-annual variability results in acute water crisis and high prevalence of surface water use by communities and schools; and (d) Lodwar town, a rapidly urbanising small town in the parched drylands of northwest Kenya, where the existing water supply utility grapples to meet the water demands of a growing population on the banks of a seasonal river with shocks from annual flash floods. We write about each of these sites in Chapters 2–5 of this book.

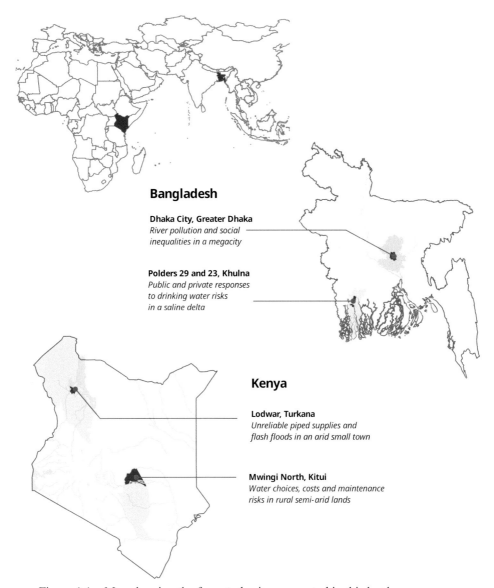

Bangladesh

Dhaka City, Greater Dhaka
River pollution and social inequalities in a megacity

Polders 29 and 23, Khulna
Public and private responses to drinking water risks in a saline delta

Kenya

Lodwar, Turkana
Unreliable piped supplies and flash floods in an arid small town

Mwingi North, Kitui
Water choices, costs and maintenance risks in rural semi-arid lands

Figure 1.1   Map showing the four study sites presented in this book.

The daily water diaries inherently captured the seasonality of water risks and practices at household and individual levels, with the design being adapted to the risks and contexts being studied. The water diaries in Dhaka were different to the ones in the other three sites, as the aim was to observe how different people interact with the rivers depending on place, time of the day, and season of the year. The 'river diaries' involved direct observation of river use behaviour that were recorded by enumerators stationed at 6 selected points along the riverbanks for

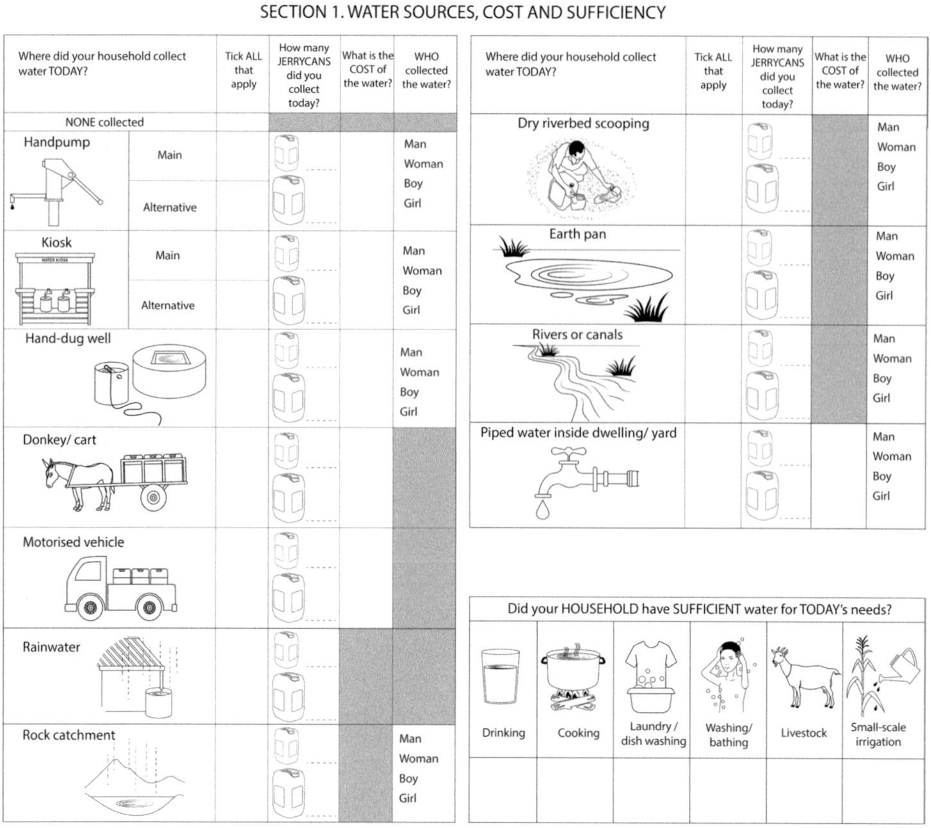

Figure 1.2   The water diary charts designed for Kitui county, Kenya, which were translated to the local language Kikamba. One hundred and fifteen households in Kyuso and Tseikuru wards of Mwingi-North subcounty participated in the diary study from August 2018 to July 2019.

33-days over two seasons – the dry season in January–February and the wet season in August–September 2019. Enumerators noted down who were using the river (that is, adults/children, male/female) for what activities over a 9-hour schedule each day, resulting in more than 10,000 observations with photographs. The water diaries in Khulna, Kitui, and Lodwar focused on drinking water services, studying 120, 115, and 98 households, respectively, for 52 weeks in 2018–2019. The design involved pictorial charts where a member of the household recorded their source, amount and cost of water for that day, along with their overall household expenses (Figure 1.2). Appendix outlines the methodological details of *The Water Diaries*, including the design, piloting, sampling, and experiences from the field.

The diaries were complemented with a suite of methods to understand the environmental, institutional and infrastructure risks shaping the water insecurities in these sites (refer to Appendix). Household surveys were conducted in each site to understand the demographic and socioeconomic profiles, state of water and sanitation facilities, and how water-related challenges rank within other development concerns. The survey households provided the sampling frame for the water diaries in Khulna, Kitui and Khulna, with participants being selected across the spectrum of reported water concerns and welfare. Hydrogeological analysis and climate models sketched the environmental risks related to groundwater and surface water quality, rainfall variability, and climate change. Water infrastructure audits recorded the locations, functionality, investments, and management of different water supply technologies. In-depth interviews with diary participants, riverbank residents, and water point managers provided detailed insights into decision-making processes for navigating water risks.

1.3 Risks, Inequalities, and Policy Responses in Bangladesh and Kenya

Through the diaries and the complementary methods, we explore the social, spatial, and seasonal dynamics of individual and household water practices, in the context of diverse water risks across geographies and sociocultural environments. People's choices and behaviour reveal the 'acceptable' level of risks, as governed by their embodied habits and the wider environmental, infrastructure, and institutional context. We explore how policy responses, addressing institutional, information and investment gaps, can better allocate responsibilities between various public and private actors to manage these risks and reduce inequalities.

We start our narrative with Dhaka (Chapter 2) where river pollution by textiles and leather industries, coupled with sewage and solid waste disposal, has severed the once close-knit bond between the city's people and waterways. Regulatory non-compliance has become normalised as successive military and democratic regimes post-independence have favoured export-oriented economic growth, in line with neoliberal policies prescribed by international financial institutions. Lax enforcement of environmental laws by the state has spurred private governance initiatives by global fashion brands and civil society, with several billion-dollar projects leading to a decentred regulatory framework (Peters, 2022). Water quality monitoring is thwarted by data gaps stemming from infrequent sampling and limited coverage that do not capture the pollution dynamics in the factories or the rivers.

In Chapter 2, we present the seasonal changes in the state of river health across different stretches using data from the first water quality monitoring system for the entire Greater Dhaka watershed. Religious events – the annual Bishwa Ijtema and Eid-ul-Azha – add pollution shocks to the system. Our river water diaries capture the social inequalities in pollution exposure by analysing the location, season and gender disaggregated interactions with the river. Individual practices by low-income riverbank communities reflect their habitus, shaped by past relations with cleaner waterbodies both in Dhaka and in rural areas where many have migrated from. Given the generational timeframes required to achieve river restoration masterplans, there is a moral obligation to take immediate actions to protect these vulnerable people through better water and sanitation facilities and risk communication. This can deter use of river for daily washing, bathing, and irrigation purposes, even when the quality is perceived to be better in monsoon.

We next move to Khulna in the coastal floodplains (Chapter 3), where the waterscape is dominated by mighty tidal rivers interlacing embanked islands. Rural populations obtain their drinking water from tube wells of varying depth. Despite high coverage of tube well – the low-cost improved technology popularised to curb diarrhoeal risks from surface water – access to safe water all year round is compromised by high groundwater salinity. Publicly financed and community managed tube wells have been the dominant approach for rural water supply, leaving out areas with high salinity that require alternative water supply technologies. Uncoordinated investments by donor organisations, households and small enterprises have emerged to plug the gaps through rainwater harvesting systems, small piped schemes, reverse osmosis plants, and pond sand filters.

Household water diaries reflect four behavioural clusters, characterised by commonalities in source choices and expenditures driven by rainfall and local salinity risks. Yet within clusters, individual preferences and habits often explain divergent practices. Uncertain water quality risks from multiple sources, infrastructure operation and maintenance risks, and financial risks related to coping costs jeopardise water security not only in domestic settings, but also in schools and healthcare centres. In this chapter, we advocate for shifts towards professional maintenance services that we piloted in public schools and community health centres in selected unions and later upscaled to the entire district through results-based financing from government and donors. Through regular water quality monitoring, prompt repair and preventative maintenance, supported by up to date information systems and regulatory oversight, professional service delivery has potential to address the long-standing functionality and water safety challenges.

From Bangladesh's water-rich delta, we move to the semi-arid hinterlands of rural Kenya (Chapter 4). In 2016, our study site Kitui emerged as 1 of 47 devolved counties with responsibility for the 2010 constitutional commitment of safe drinking water to all Kenyans. By the 2019 national census, two in five Kitui residents stated their main source of drinking water was surface water, compared to the national average of 23 per cent. This a remarkable statistic given the economic status of Kenya and the investments made over many decades to improve drinking water services by national government, bilateral donors, and non-government organisations (NGOs). Water diaries show shifts from groundwater to surface water sources immediately after rainfall reflecting the cultural predilection towards free and freshly collected waters for people and livestock. The seasonal transition between water supplies reduces revenues for operation and maintenance of piped schemes and handpumps, further aggravating the chronic functionality challenges of community managed systems. While the water sector act explicitly encourages counties to contract private entities to address financing and operational challenges, the commercial non-viability of water services in sparsely populated areas with inconsistent demands deters private sector participation. County leadership and donor cooperation can spur uptake of professional maintenance service delivery guaranteeing reliable and safe drinking water services.

Our final destination is one of the driest inhabited places on the planet – the small town of Lodwar (Chapter 5) in Kenya's northwestern Turkana County bordering Ethiopia, South Sudan, and Uganda. Turkana is famed for being the cradle of humankind with discoveries of the Turkana Boy and Lucy by archaeologists. Lake Turkana is the largest of the great soda lakes of the Rift Valley and marks the terminus of the river Omo, originating in southern Ethiopia, and the Turkwel River flowing from the highlands of southern Uganda. Turkana is an unforgiving dryland with very low rainfall partly due to low-level and high-speed air currents, which transport vast counties of vapour from the Indian Ocean to the Congo basin in the west, leaving Turkana unnaturally dry. The recent discovery of oil deposits and large groundwater reserves has generated excitement and investment in a region purposively marginalised under British colonial rule. In parallel, as a geographic anchor linking the hinterlands of three countries, there has been an ambitious infrastructure project connecting Lodwar with the Lamu Port – South Sudan – Ethiopia Transport corridor project (LAPSSET).

Our diaries reflect the daily water challenges of this rapidly growing town against the backdrop of harsh environment, historical marginalisation, and high poverty. Infrastructure and institutional inefficiencies have resulted in an unreliable piped water service with limited coverage, providing a fertile ground for informal water markets to flourish particularly in unserved peri-urban localities. Those living by the rivers face the dilemma of migrating to remote areas without schools, employment opportunities or water services, or living in fear of being washed away by flash floods. Groundwater is a strategic economic resource in these drylands, with multiple demands from urbanisation, irrigation, and oil exploration. Groundwater sustainability requires knowledge of recharge processes and hydrochemical characteristics to be monitored and managed effectively by government.

The water diaries of urban and rural populations across these four sites illustrate the local experiences of global water risks. We do not intend to prescribe solutions, rather critically reflect on the complexities of water-society dynamics to guide public and private responses. The climate crisis is neither the unique cause nor the singular solution to the global water crisis. However, it provides an important political and funding framework where appropriate and effective action to mitigate and adapt would deliver significant benefits to vulnerable people. With the mid-term evaluation of progress on the SDGs in 2023 providing a bleak summary for the water goal and the other 16 goals, there is an opportunity and obligation to think of alternative actions and behaviours to a shared commitment to improve and maintain water security in a rapidly changing and unstable world.

2

Fashionable Rivers

Social Inequalities and Pollution in Dhaka

2.1 Introduction

Dhaka, home to an urban population of 20.7 million (BBS, 2023), is amongst the world's most densely populated places on the planet. Despite being infamous for its traffic congestion and river pollution, this 400-year-old city was once fondly referred to as the 'Venice of the East', with its vibrant rivers and canals defining the city's trade, transport, and sociocultural identity. From the seventeenth-century capital of the Mughal Empire, celebrated for its high-quality yet cheap muslin and silk exports, Dhaka has transformed into a global manufacturing hub for leading fashion brands, driving the country's rapid economic growth in the twenty-first century. With 'Made in Bangladesh' labels increasingly making their way into European high streets and wardrobes, the colourful discharge from thousands of factories silently taints Dhaka's rivers and waterbodies. Millions of litres of untreated municipal wastewater and solid wastes are further added to this daily cocktail, making the rivers the backdoor drains that have little connection with the city's middle- and upper-class residents. The riverfronts that were once sought-after residential areas are now occupied by the most impoverished citizens.

What we see in Dhaka today reflects a modern iteration of the historical 'Great Stinks' that afflicted rapidly industrialising European cities in the nineteenth century. London, for instance, faced a sanitation crisis where faecal waste from three million residents was routinely dumped into the River Thames, leading to frequent cholera outbreaks (Halliday, 2001). The tipping point came during the scorching heatwave of the summer of 1858, when an intolerable stench pervaded the Victorian city. This prompted the construction of London's state-of-the-art sewage system over the subsequent decade. Today, high-income countries in Europe and North America treat 70 per cent of their wastewater before discharge, compared to only 8 per cent and 28 per cent in low- and lower-middle-income countries in Asia, respectively (Kookana et al., 2020, WWDR, 2017, UNEP, 2016).

Improvement in ambient water quality by halving the proportion of untreated wastewater by 2030 is one of the targets under the Sustainable Development Goals (SDG 6.3). However, monitoring global and national progress towards SDG 6.3 is hindered by substantial data gaps, mainly due to the absence of systematic water quality monitoring systems in most developing countries (UNEP, 2021). The existing water quality monitoring system for Greater Dhaka has insufficient coverage, frequency, and parameters to allow a system-wide understanding of pollution dynamics and effectiveness of interventions. High organic pollution from sewage and solid wastes can interact with chemicals in industrial effluent, amplifying overall toxicity for living organisms. During the monsoon season, runoff from densely populated urban areas and agricultural lands significantly contributes to the pollution load. The heightened river flows and floods transport pollutants downstream and into the floodplains of rural areas. In this chapter, we closely examine these pollution dynamics through monthly data on organic, inorganic, heavy metal, and pathogen contamination across the six rivers surrounding Dhaka.

It is difficult to evaluate and attribute the risks of river pollution to human health and well-being due to the multiple exposure pathways, the long latency periods, and the undefined spatial scales of observable impacts. Beyond direct health effects such as gastroenteritis and skin diseases from contact or immersion in polluted water (Turbow et al., 2003, Prüss, 1998) and consumption of heavy metals accumulated in fish and crops (Khan et al., 2008, Wang et al., 2005), there are broader impacts on well-being stemming from poor visual amenities, unpleasant odours, limited recreational opportunities, and stigmatisation of communities (Damery et al., 2008). This led us to study the 'river water diaries' of Dhaka to understand who interacts with the rivers, for what purpose, and when. Shifting the focus from observable outcomes, this approach allowed a nuanced understanding of the social, economic, and cultural significance of the rivers in people's lives across space and seasons.

We situate these risks and inequalities within the historical development trajectory of Dhaka from the Mughal era to the British colonial period and discuss how post-independence political and economic priorities shaped the discourse of environmental regulations. With Bangladesh's economic ascent, transitioning from a 'least developed country' to a 'lower-middle-income country' in 2015 and set to achieve a 'middle-income country' status by 2026, there is an increased political will to free the rivers from pollution. However, this Herculean task is complicated by the complex network of stakeholders within the government, civil society, and the global textile industry, with various forms of state interventions and market-led governance operating within a messy regulatory space. In this chapter, we portray Dhaka's fashionable rivers to paint the complexities of water–society dynamics that challenge sustainable actions to mitigate risks. The daily experiences of pollution risks by marginalised populations in a megacity reveal one of the many faces of the global water crisis.

2.2 Connected by Rivers, Disconnected from Rivers

In March 2018, on a hot and humid morning, we gathered by the banks of the Buriganga River, awaiting to board a boat for a river tour through the waterways of Dhaka. With a team of about 50 national and international academics and water sector practitioners, the trip was intended to get a first-hand experience of the unfortunate state of pollution of Dhaka's mighty rivers. Despite being labelled as 'ecologically dead', the bustling ambience of the Buriganga waterfront was a testament of the river's historical role in sustaining the city's commerce and communication. Multi-tiered ferries, commonly referred to as 'launches', were poised to embark on journeys to the southern delta – a landscape interwoven with rivers and creeks that culminate into the expansive Bay of Bengal (Figure 2.1). Cargo freighters with construction materials and motorised wooden trawlers with colourful local produce were docked, waiting to be offloaded. Soon porters lined up to transfer the produce to the vans, like a human conveyor adeptly balancing the loaded baskets on their heads and seamlessly switching it with the next person in exchange of an empty one.

Figure 2.1   Looking into the expansive Buriganga River from our tour boat in March 2018. As the black, polluted waters glistened in the morning sun, the bridge at the far end read, 'The nation thrives if the rivers survive. We will bring back our Golden Bengal' (translated from Bangla). (Photo credit: Alice Chautard, 2018).

This is the birthplace of Dhaka – the northern banks of the Buriganga River where the Mughals established their provincial capital in the early seventeenth century. Cradled by six rivers like a garland – Buriganga, Dhaleswari, Turag, Tongi Khal, Balu, and Sitalakhya Rivers – Dhaka's strategic location was of both military and commercial interest. On the one hand, it enabled surveillance of the lower Bengal Delta against enemy attacks, while on the other hand, the network of rivers and canals allowed transportation of goods to the inner empire. Writing about the seventeenth-century Dhaka, English mariner Thomas Bowrey referred to it as a 'large spacious' metropolis, with 'very fine conveniences' as it stood beside a 'fine and large river' navigable for ships of 500–600 tons (Bowrey, 1905, p. 150). The water of the Buriganga 'being an arm of the Ganges' was 'extraordinarily good' (Bowrey, 1905, p. 150), serving drinking water to its one million dwellers. French merchant Tavernier further noted that Dhaka extended only in length, as 'everyone coveted to have a house by the Ganges-side' (Tavernier et al., 1684).

Today, the Buriganga River, along with the interconnected river system of Greater Dhaka, is amongst the top contenders of the world's most polluted rivers. This was apparent as we sailed north-westwards along the Buriganga towards the Turag. The black and indigo water, the rotten egg-like stench, and organic and plastic wastes littered along the banks were in sharp contrast to the picturesque waterscape narrated by Bowrey and Tavernier. The silver jewellery of one of our colleagues became tarnished from silver to black, potentially from the hydrogen sulphide in the air. The appalling state of the rivers is the result of the indiscriminate discharge of untreated wastewater from the city's households and manufacturing industries, mainly textiles and tanneries. The factories are located within major planned and unplanned industrial clusters along the riverbanks (Figure 2.2). The highest pollution burden is imposed by the wet processing textile units, owing to the high usage of water, along with salts, dyes, and bleaches for washing, dyeing, and finishing processes. Apart from the larger textile units within the export processing zones, most small and medium factory units are concentrated in informal, heterogeneous, under-serviced industrial clusters, often interspersed with some residential dwellings. With very few textile units equipped with fully functioning effluent treatment plants, the bulk of the untreated wastewater makes their way into the rivers through bypass drains and internal canals.

Compared to textiles, tannery wastewater comprises a high concentration of biodegradable organic load and a substantial nitrogen content, given that leather is composed of proteins, keratins, and fats. The tanning process involves the use of large amounts of salts to preserve the leather, resulting in elevated levels of alkalinity and inorganic compounds, including chromium, chlorides, ammonium, sulphides, and sulphates (Sagris and Abbott, 2015). Even though the total volume of wastewater generated by the tanneries is considerably lower than that of the textile industry, the

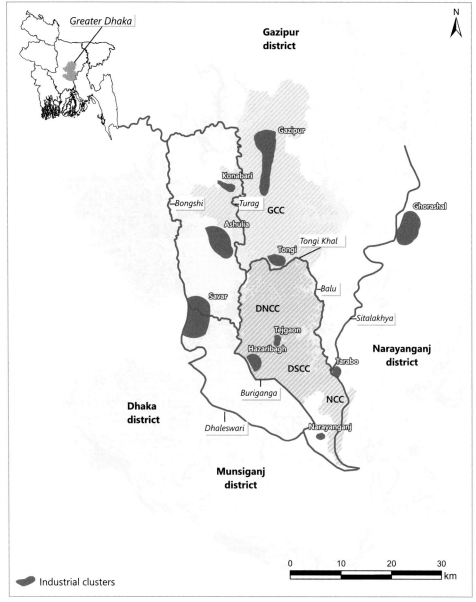

Figure 2.2   Map of Greater Dhaka (comprising four districts) showing the Dhaka North (DNCC), Dhaka South (DSCC), Gazipur (GCC), and Narayanganj (NCC) City Corporations, major rivers, and industrial clusters.

demise of the Buriganga River can be attributed to several decades of pollution stemming from the tanneries in the Hazaribagh cluster (Whitehead et al., 2019). With the tannery industry now been relocated to Savar, the Buriganga River is showing gradual signs of recovery at the cost of the Dhaleswari River now being contaminated.

Today, Bangladesh stands as the world's second-largest exporter in the USD 45 billion ready-made garment industry, trailing only China, and commands an 8 per cent share of the global market (Berg et al., 2021). With this sector contributing over 80 per cent of the country's export revenue and employing 4 million people, it has emerged as the primary driver for Bangladesh's economic growth since the 1990s. The leather industry, which contributes around 3 per cent to the export revenue, is also of strategic importance as the country is seeking to diversify its manufacturing export base to sustain the growth trajectory (Hong, 2018).

Dhaka's status as a textile exporter, however, can be traced back to the Mughal period. In contrast to the present-day exploitative dynamic, where the global demand for cheap and fast fashion is choking the local river system, textiles and rivers once shared a harmonious relationship. The exquisite muslin fabric, sought after by royalty and traded across the Roman and Ottoman empires during the sixteenth and seventeenth centuries, was handwoven by skilled artisans using the finest cotton grown along the rivers (Islam, 2016). The thread was spun in intensely humid conditions, typically in the morning and evening, with young women going to the middle of the river by boat to cut the yarn (Gorvett, 2021). The flourishing muslin trade in Bengal, however, met its demise at the hands of the British colonial machinery, whether wielded through the East India Company or direct rule by the Crown. This downfall was orchestrated through discriminatory taxes and tariffs against local cotton growers and weavers, favouring machine-produced imports from British cotton mills.

The colonial policies of the nineteenth and early twentieth centuries also shifted the focus of development from water to land. The dynamic landscape of the delta – its shifting rivers, monsoon flooding, and accretion of new lands – was perceived as a hindrance in the functioning of territorialisation, governance, and taxation. A discourse of 'contained water' and 'dry ground' emerged, gradually eroding the close-knit bond between Dhaka and its rivers. Under British influence, the transformation of medieval Dhaka into a modern city unfolded, marked by the construction of metalled roads, open green spaces, streetlights, and piped water supply (Ahmed, 1986). The first flood-protection embankment along the Buriganga was also erected. The advent of the railway in the late nineteenth century further marginalised the significance and upkeep of rivers and internal canals as commercial arteries. The colonial legacy persists in shaping the city's development trajectory, as wetlands, canals, and rivers continue to be encroached upon for real estate development (Baffoe and Roy, 2023). For Dhaka's middle and upper classes today, encounters with rivers are limited to media coverage of pollution issues, highlighting a stark disconnection from the once integral and dynamic relationship between the city and its waterways.

2.3 Monitoring River Health

The existing institutional arrangement for monitoring surface water quality in Greater Dhaka involves monthly measurements of selected physiochemical parameters by the government's Department of Environment (DoE), though data collection is less frequent in practice (DoE, 2017).[1] The DoE's network of 17 sampling points does not cover the rapidly urbanising areas along upper Turag and Balu Rivers, as well as some important industrial zones. Assessment of heavy metals, inorganic parameters, and persistent organic pollutants in water and floodplain sediments is also needed for a comprehensive understanding of pollution dynamics at the system level. This involves understanding the relative contribution of pollution loads from different reaches and old sediment deposits and the biochemical interactions of pollutants that affect overall toxicity and evaluating the effectiveness of ongoing and planned interventions.

A water quality monitoring system, developed by our colleagues at the Bangladesh University of Engineering and Technology (BUET) (REACH Dhaka, 2023), is the first attempt to ensure regular monitoring of river health across Greater Dhaka. Monthly data from 58 sampling points show the seasonal changes in water quality across different stretches (Figure 2.3). Details on data collection and analysis are outlined in Appendix. During low flow conditions, that prevail from December to April, the dissolved oxygen throughout the river system falls below the minimum threshold essential for sustaining aquatic life. The anoxic conditions in the waters and sediments can result in some metals from bound sediments and gases such as methane, ammonia, and hydrogen sulphide being released. Among the most polluted stretches are the 29-km Buriganga River in the south-west and the Tongi Khal, a 15-km canal linking the Turag and Balu Rivers in northern Dhaka.

The Tongi Khal is also known for its religious significance. Since 1967, its banks have been the sacred grounds for the Bishwa Ijtema, the second-largest gathering of Muslims worldwide, following the Hajj pilgrimage. Every January, a 160-acre government land is transformed with tents to host three million devotees over two three-day phases. The Ijtema served as a natural experiment to understand the impacts of human activities on river water quality. An analysis of heavy metals in Tongi Khal before and after the Ijtema showed an overall increase in concentrations between December 2017 and January 2018 (Rampley et al., 2019). This was due to the combined effects of decreased river flow, resuspension of sediments caused by the disturbance of riverbed during Ijtema preparatory work, and subsequent dissolution of metals due to anaerobic condition. The Ijtema event

---

[1] Existing physiochemical parameters monitored by the Department of Environment (DoE) include pH, electrical conductivity, total dissolved solids, suspended solids, dissolved oxygen, alkalinity, biological oxygen demand, chemical oxygen demand, chloride, total coliform, and *Escherichia coli* (*E. coli*).

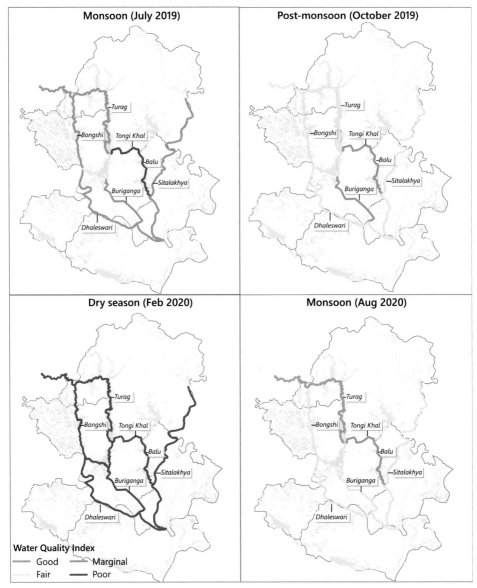

Figure 2.3   River health of Greater Dhaka during 2019–2020 based on a Water Quality Index comprising 15 parameters, namely temperature, pH, electrical conductivity, dissolved oxygen, oxidation-reduction potential, turbidity, colour, alkalinity, iron, ammonia nitrogen, nitrate, phosphate, sulphide, sulphate, and chloride. Drawn by author using data collected by the Bangladesh University of Engineering and Technology under the REACH Programme.

itself introduced a variety of heavy metals into the Tongi Khal. Washing utensils contributed aluminium, chromium, and iron, while the likely consumption of medicines added selenium. Improper disposal of batteries introduced lithium into the water, and fish residues released manganese. A bacterial biosensor was developed to understand the toxicity of these combinations of heavy metals on living organisms (Rampley et al., 2019). Interestingly, despite the overall increase in metal concentrations, the simultaneous dumping of organic waste resulted in lower availability of free ions, which in turn lowered the toxicity in the samples collected during the Ijtema.

The onset of the monsoon in June–July transforms the riverine landscape. The combined effect of upstream flows from the Himalayas and localised rainfall within the basin leads to the rivers rising (Figure 2.4), subsequently flooding the adjacent low-lying regions. The dilution fades away the stench and black mirror-like surface of the rivers, indicating recovery of the river health to some extent. The concentrations of most organic and inorganic pollutions fall below national thresholds, while the dissolved oxygen rises (Hoque et al., 2021). There is, however, significant local storm runoff from urban and industrial areas into the streams and rivers, which is evident from elevated turbidity levels in rivers throughout the monsoon season. Increase in temperature and dissolved oxygen, coupled with addition of different genera of coliform bacteria from the monsoon runoff, also reduces the relative proportion of faecal coliform (*E. coli*) to total coliform.

Interestingly, the best condition is observed in the post-monsoon season as flushing of the floodplains and rivers in monsoon results in lower sediment and cleaner water, with the pollution plume travelling further downstream (Figure 2.3). The extent of monsoon flooding also varies from year to year. The 2020 floods, for example, inundated about 18 per cent of Dhaka district and exposed 3 per cent of its population – twice as much as 2019 floods (Khan et al., 2024). Moreover, in 2020 as factories were temporarily shut due to COVID-19 lockdown, overall river health was substantially better compared to other years. While monsoon flows

Figure 2.4   Low-income settlements near the Konabari industrial cluster along the Turag River. Rise in water levels in monsoon dilutes pollutants, though increased proximity and use of river water is likely to increase exposure to toxic chemicals and pathogens. (Photo credit: Sonia Hoque, February and August 2019).

dilute the pollutants, the floodwaters deposit toxic heavy metals on the agricultural land affecting dry-season crop productivity. This issue is, however, understudied owing to the widespread perception of monsoons being a blessing as they bring in fertile alluvial soils.

2.4 Life on the Banks of Dead Rivers

Despite the pollution, the banks of the rivers, especially those traversing dense urbanised areas, are always busy with people engaged in diverse activities. We first visited the Tongi slum on a foggy winter morning in January 2017. As we walked along the narrow passages, littered with domestic waste, the honking of buses and trucks on the road faded a little, while the stench of the polluted water grew stronger. Upon reaching the riverbank, we found women and girls washing clothes, dishes or cleaning raw fish and vegetables in the pitch-black river water. A hundred meters away were two hanging latrines. A number of boats were also docked along the bank, being home to the 'bede' ethnic group who live and travel along the rivers. Across the river, close to a major fish market, we saw men washing bamboo baskets used to carry or store fish.

Later that year, we started to survey about 2,000 households along a 25-km stretch of the Turag River and Tongi Khal with a view to study the lives of marginalised communities living close to the riverbanks (Hoque et al., 2021). We divided our study area into four zones, covering up to 1 km on both sides of the riverbank (Figure 2.5). Zone-1, on the upper reaches of the Turag River, comprises land privately owned by a few families descending from early settlers, as well as long-term leaseholders, who built and rented out houses to families and individuals working in nearby brick kilns and factories. Zone-2 covers six peri-urban settlements along the lower reaches of the Turag River, while Zone-3 comprises three densely populated slums close to the Tongi industrial cluster along the Tongi Khal. These slums, including the one we visited above, are on government-owned land, with most tenants and homeowners lacking legal ownership, putting them at risk of eviction. Zone-4, the final stretch of the Tongi Khal, is a relatively quieter neighbourhood, with a boat terminal on the southern bank.

While the surveyed households were generally amongst the poorest residents of the city, there were marked differences in socio-economic profiles of the different zones. Poverty was highest in Zone-3, with one in five adult women working in one of the factories nearby, while the men were engaged in small businesses, casual labour, and garment factories. Zone-4 was a relatively well-off area, having the highest proportion of adults working in the service sector. The average household monthly expenditure in Zone-4 was USD 300, which was higher than the other zones with mean expenditures ranging from USD 160 to 200.

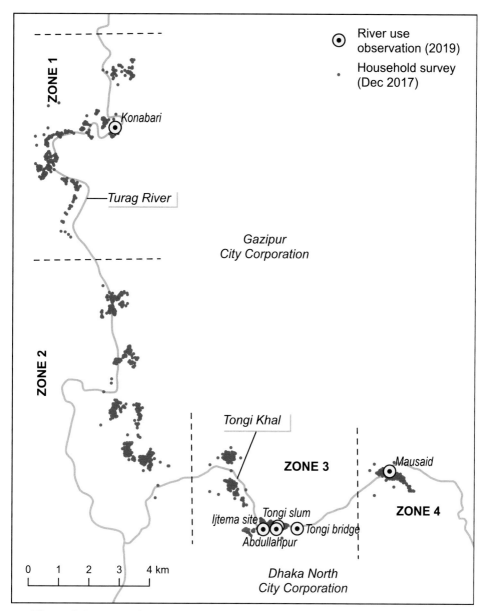

Figure 2.5 Map of Turag River and Tongi Khal in northern Dhaka showing locations of households surveyed and river use observation study by zones.

Likewise, there were differences in access to water and sanitation facilities across the four zones (Table 2.1). Households in Zone-1 and Zone-2 used motorised tube wells installed by NGOs, either through pipelines drawn inside their dwelling or through shared taps in communal spaces. Those living on the southern bank of the Tongi Khal in Zone-3 had access to piped water from the water utility (Dhaka

Table 2.1   *Water and sanitation facilities of households along Turag River and Tongi Khal.*

Household characteristics		Zone-1 (%)	Zone-2 (%)	Zone-3 (%)	Zone-4 (%)
Wealth quartiles	1 (Poorest)	23	18	41	4
	2	26	28	28	11
	3	30	26	20	18
	4 (Richest)	21	28	11	66
Main source of drinking water	Piped water into dwelling/yard	20	6	55	1
	Motorised tube well	73	92	42	96
	Others	7	2	2	3
Water source sharing	Not shared	44	26	11	75
	Less than 5 households	33	33	21	20
	5–10 households	14	15	27	5
	More than 10 households	9	26	41	0
Sanitation	Flush to septic tank	19	41	26	79
	Improved pit latrine	68	45	43	20
	Unimproved pit latrine	3	2	5	1
	Hanging toilet/Open defecation	10	12	27	1
Toilet sharing	Not shared	61	47	20	86
	Less than 5 households	27	25	33	13
	5–10 households	9	9	25	1
	More than 10 households	4	19	22	0
General concerns	Clean environment	3	3	25	11
	Water supply	6	8	17	0
	Sanitation	3	3	16	1
	Healthcare	23	32	7	6
	No concerns	1	2	13	9
	Others	23	42	23	42

Note: Data from 1,826 households surveyed in December 2017.

Water Supply and Sewerage Authority (DWASA)), whether legally or illegally, while people on the northern bank relied on community tube wells. While they had access to an improved source within a few minutes of walking distance, water was often only available at certain times during the day, creating long queues for a limited quantity that is prioritised for drinking and cooking. Two-thirds of the households in Zone-3 reported sharing their water source with at least five other households. In contrast, those in Zone-4 usually had their own tube wells, with a quarter of them sharing it with others in the same compound.

Although none of the households were connected to formal sewerage infrastructure, those in Zone-4 used a flush toilet with a septic tank that was not shared with others. The sanitation conditions were worst in Zone-3, with more than a quarter using hanging latrines or practising open defecation, with the faecal waste discharging into the river. About half of the households in Zone-3 shared their toilets with at least five households, suggesting the lower living standards and increased likelihood of spreading pathogens. Improved pit latrines were more common in Zone-1 and Zone-2, followed by flush toilets with septic tanks.

These differences in poverty and living standards were closely linked to exposure to river pollution. To understand who interacts with the river, under what circumstances and for what purposes, we conducted a direct observation study for two weeks in the dry and wet seasons, respectively. Six observation sites were selected, four in Zone-3 and one in Zone-1 and Zone-4, where enumerators were stationed for three-hour slots throughout the day. Zone-2 was excluded as we did not find any interactions with the river during our scoping visits, as the short river branch flowing through the bottom part of Zone-2 remains dry for part of the year.

Direct contact with river water, through dishwashing, laundry, cleaning fish and vegetables, and personal washing was high in Zone-1 and Tongi slum in Zone-3, particularly among women and girls (Figures 2.6 and 2.7). These sites were close to residential clusters, where overcrowding at community water points and restricted supply only at certain times during the day, led to high usage of river water for domestic activities. Domestic activities were relatively fewer in Zone-4, owing to the better socio-economic status of residents in this area. During the wet season, we observed a slight decrease in domestic activities in Zone-1 and Tongi slum in Zone-3. In Zone-1, construction of new houses along the bank which partially deterred the local residents from accessing one of the observation sites, while in Zone-3, around 60 houses were evicted as part of the government's drive to control river encroachment.

Small-scale productive uses were also observed at some sites in Zone-3. These included washing and dyeing denim, washing fish baskets or plastic sheets, collecting plastic waste, and fishing. Denim washing was mostly carried out by men, under the pillars of the railway bridge on the Tongi Khal. Informal waste pickers could be spotted wading through the river on a boat, collecting plastic bottles for resale while others were seen washing plastic sheets either on the banks or in waist-deep waters. These indigo tainted sheets are waste products from dye packaging, which serve as an income source for these marginalised citizens. Fishing, with or without a boat, was commonly observed across all sites during the wet

Figure 2.6 (a) Boat dwellers and hanging latrines along Tongi slum (Photo credit: Sonia Hoque, 2017); (b) woman collecting plastic bottles from Tongi Khal (Photo credit: Rebecca Peters, 2017); (c) men washing and bathing in the indigo waters (Photo credit: Alice Chautard, 2018); (d) and a Ferris wheel for children next to an effluent outlet along Buriganga River (Photo credit: Alice Chautard, 2018).

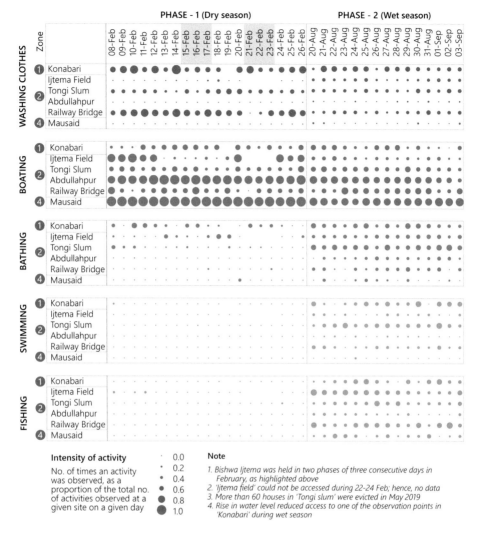

Figure 2.7   Intensity of river use activities disaggregated by zone and observation site. Reprinted from Hoque et al. (2021) under the terms of the CC BY 4.0 license.

season. In Zone-1, the abundance of fish along the banks meant that women and children could easily catch these with their bare hands.

Monsoon brought about a steep increase in swimming and bathing activities in Zone-1 and Tongi slum in Zone-3. Recreational swimming was more prevalent among adult men and male children, as cultural norms sometimes refrained women from sharing the same public space with men for recreation. The rise in bathing and swimming, which was also prevalent in other sites, can be attributed to the warmer weather and perceived improvement in water quality. The river was heavily used for boating, mainly for transportation and sometimes for recreation. Boating increased significantly in Zone-3 and Zone-4 during the Ijtema period for transportation of people, food, and construction materials. While temporary sanitation facilities with faecal sludge containment was provided for the Ijtema devotees, urinating into the river was commonly practised, along with the disposal of organic food waste.

Although monsoon flooding typically rejuvenates the Bengal Delta by depositing nutrient-rich alluvial sediment, the scenario is different in Dhaka. Here, the monsoon also brings in toxic pollutants closer to settlements and croplands. The relocation of the tanneries from Hazaribagh to Savar, which we elaborate further in this chapter, has not only transferred pollution from the Buriganga to the Dhaleswari River but has also unleashed new challenges on the downstream communities, particularly impacting Hazratpur village situated 5 km downstream of the Savar tannery estate. Hazratpur, an agrarian community, cultivates rice from December to July, with the low-lying croplands submerged during the subsequent months. The cultivation of vegetables, especially carrots, occurs on higher ground in three-month cycles throughout the year, proving to be a more lucrative output. However, over the past five years, the untreated tannery waste has taken a toll on soil fertility and led to a surge in insect infestation. The pollution load reaches its peak a few weeks after Eid-ul-Adha,[2] coinciding with flood timings in recent years. Residents of Hazratpur also expressed deep concerns about the declining groundwater level in the village, attributed to the recent installation of five production boreholes by DWASA. This village, once water-secure with access to freshwater through deep tube wells for drinking and irrigation, now grapples with adversities arising from the city's population growth and industrial activities.

2.5 Regulation by State, Market, and Civil Society

Amongst the various pollution sources, the textile industry often garners the most scrutiny owing to its economic and political significance for Bangladesh, as well as well-known clothing brands. Photos of frothy indigo effluent gushing into the rivers, with boats navigating through floating plastic wastes, appear in international news media echoing the impacts of fast fashion on the poorest populations. Yet, despite compulsory regulations for effluent treatment, along with increasing emphasis on environmental sustainability from global fashion brands, river pollution continues unabated. Understanding this normalisation of pollution and regulatory non-compliance requires critical inquiry into the evolution of state-market dynamics since the country's independence in 1971 (Peters, 2022).

---

[2] Eid-ul-Adha, a Muslim festival observed on the 10th day of the last month of the lunar Islamic calendar, shifts approximately 10 days earlier in the Gregorian calendar each year. In Bangladesh, this festival involves the ritualistic sacrifice of cows and goats, resulting in a substantial supply of raw hides.

Bangladesh's early years were marred with economic and political instability. The severe famine of 1974, followed by the assassination of Prime Minister Sheikh Mujibur Rahman in 1975, significantly hindered the country's efforts towards post-war recovery. In contrast to Mujib's democratic-socialist ideology, marked by nationalisation of industrial assets and a highly regulated financial sector (Islam, 1985), the successive military regimes under General Ziaur Rahman (1977–1981) and General Ershad (1983–1990) adopted a capitalist liberal economic approach. With the promulgation of the New Industrial Policy in 1981, further reformed in 1986 and 1992, the government gradually retreated from its 'regulatory' role to facilitate the development of the private sector and expand the country's export base to non-traditional sectors such as garments and frozen fish. Through a compendium of trade and fiscal reforms, including concessionary duties for imported raw materials and machineries, tax rebates on export income, deregulation of interest rates, reduced emphasis on loan recovery, and a flexible exchange rate with frequent depreciation, the private sector assumed a greater control of driving the country's economy (Mohammad and Alauddin, 2005).

Although these macroeconomic reforms were presented as initiatives to promote human and economic development, they essentially served as a tool for Zia's and Ershad's military regimes to legitimise and strengthen their unconstitutional power structures by forming alliances with senior bureaucrats and industrialists (Quadir, 2000). During Khaleda Zia's democratic regime (1992–1997), the private sector's influence grew even stronger. A significant majority of Parliament members were businessmen and industrialists who funded the party's extravagant electoral expenses. These policies of market liberalisation were also embraced by international financial institutions such as the World Bank and the International Monetary Fund, which gained unprecedented influence in guiding national policy reforms in aid-dependent countries in the global south (Sobhan, 1993). These broader economic and political conditions provided a fertile ground for Bangladesh's textile industry to flourish during the 1980s. At the same time, Bangladesh's initial exemption from the Multi-Fibre Agreement, which granted quota-free access to the United States market, attracted foreign private investments in manufacturing factories. This, in turn, facilitated significant development of skills and industry knowledge among local entrepreneurs, leading to growth of domestic ventures under the patronage of successive political regimes (Mottaleb and Sonobe, 2011).

Given the economic and political significance of the manufacturing sector, the government has traditionally adopted a 'retreatist' approach to environmental regulations for industries – one that is 'strong on paper' but 'weak in enforcement' (Peters, 2022). The Environmental Conservation Act (1995), and the subsequent Environment Conservation Rules (1997), require 'red category' industries, including fabric dyeing and tanneries, to obtain location and environmental clearance certificates and treat their wastewater to scheduled standards prior to discharge (MoEF, 1997). Revised in 2010, the Act establishes a minimum fine of BDT 50,000 (USD 450) and a maximum fine of BDT 200,000 (USD 1,800) for cases of 'discharging excessive pollutants' (MoEFCC, 2010). For repeat offenders, the penalty increases to 1,000,000 BDT (approximately USD 9,200). Yet, publicly disclosed data from 290 cases between 2011 and 2016 show that fines are often imposed arbitrarily, ranging from USD 117 for a garment factory lacking a clearance permit or an effluent treatment plant, to as much as USD 350,000 for instances of partial operation of an effluent treatment plant on the factory premises (Haque, 2017).

For many factory owners, the cost savings from not installing or operating effluent treatment plants far outweigh these penalty fines (Haque, 2017), with additional data from 2010 to 2019 showing that only half of the fines imposed were actually recovered by DoE (Peters, 2022). The limited enforcement actions taken by DoE can be attributed to the significant autonomy enjoyed by large factory owners, who can leverage their political connections to request reductions or waivers of fines, placing DoE officers at risk of facing professional consequences from higher authorities. As a result, cases that are simpler and easier to identify are often pursued, often singling out smaller enterprises that are less likely to mount effective resistance.

Against this backdrop of slack state regulation, private governance by international brands or 'buyers' has emerged as an alternative form of environmental regulation. With growing consumer pressure, international brands are increasingly focusing on environmental sustainability practices along their supply chains. Bangladesh now has the highest number of factories with the Leadership in Energy and Environmental Design certification, indicating a trend towards greening investments to boost buyer confidence and assurance of continued participation in the export market (The Daily Star, 2023). Since 2013, the International Finance Corporation led Advisory Partnership for Cleaner Textile has supported hundreds of factories in adopting cleaner production practices that reduce water consumption, wastewater discharge, energy usage, and greenhouse gas emissions (The Daily Star, 2017). Yet participation in such programmes is often limited to the larger first tier suppliers who can recoup the upfront investments through long-term savings in operation costs. For the vast majority of smaller factories, which often operate as subcontractors for larger facilities and fall outside the buyers' scrutiny, this market-driven regulatory approach proves ineffective in mitigating pollution.

Parallel to these state-market power dynamics, civil society organisations and environmental advocacy groups have emerged as influential actors in negotiating regulatory processes through public interest litigation. A notable example is the Bangladesh Environmental Lawyers Association, which, in 2003, filed a writ petition against relevant government agencies and tannery owners in Hazaribagh, demanding to stop polluting the Buriganga River (BELA v. GoB, 2003). In response, the High Court ordered the industry to relocate to a purpose-built tannery estate along the Dhaleswari River in Savar – a process that ultimately spanned one and a half decades. The centralised effluent treatment plant, a cornerstone of the relocation efforts, however, proved ineffective in mitigating pollution due to a mismatch between its treatment capacity and the actual effluent flow during the peak three-month season (Mirdha, 2023).

Another landmark verdict was delivered by the High Court in 2019, granting 'legal personhood' status to all rivers in Bangladesh and authorising the National Commission for Rivers as the legal guardian (Islam and O'Donnell, 2020). The verdict was a response to a writ petition filed by the Human Rights and Peace for Bangladesh to challenge the legality of earth-filling, encroachment, and construction of structures along the banks of the Turag River. The petition was corroborated by a 2016 report titled 'Time to Declare Turag Dead' (Ali, 2016) published in *The Daily Star* – the country's leading English newspaper.

Over the past one and half decades, multiple projects and plans have been designed and implemented by various government agencies and donor organisations to address these seemingly obvious solutions (Siddique and Rahman, 2019). Yet, despite hundreds of millions of dollars of investments, there has not been any noteworthy improvements in river water quality. For instance, the much-hyped Buriganga clean-up project, undertaken by the Bangladesh Inland Water Transport Authority (BIWTA) in 2012, proved to be futile as solid waste recovered from the riverbed and piled on riverbanks ultimately made their way back into the river during monsoon. In response to ongoing river encroachment, BIWTA has spearheaded multiple initiatives to dismantle illegal structures and remove informal settlements, including one in 2019 that evicted the Tongi slum we studied. A network of pillars, erected over the years to demarcate the river boundary, has allegedly been placed in the wrong locations, further legitimising powerful land grabbers. While BIWTA oversees the management of riverbanks, the flow of surface water falls under the jurisdiction of the Bangladesh Water Development Board. An ambitious project designed to augment the dry season flow of the Buriganga by desilting its links with Jamuna River was never completed.

2.6 Conclusion

River health has characterised the political and economic progress of many global cities. The health of rivers is a legacy bestowed on the current population to hand over in better condition to the next generation. Many of Dhaka's residents may be rightly concerned by both their recent inheritance and future legacy with high levels of river pollution due to the forces of rapid economic and demographic growth in the context of regulatory non-compliance environmental management. Water security inequalities have been exacerbated by river pollution with the most vulnerable least able to escape the social and health risks of living and working in the multiple river arteries of the city.

Observing patterns of daily behaviour across the city reveals who is in harm's way. Across the four zones we studied there are stark social inequalities between environmental and social risks in Abdullapur (Zone-3) and Mausaid (Zone-4). In the former, people are most concerned with a clean environment and basic drinking water and sanitation services. This reflects ten or more households sharing water supplies or toilets, with a high incidence of open defecation. Remember that Bangladesh is regarded as a regional success story in largely eradicating open defecation. In contrast, people living in Mausaid are more concerned with roads and healthcare reflecting limited toilet sharing and open defecation. The hydrological models clearly show river water pollution is concentrated in Abdullapur, and there is no coincidence that is where the most vulnerable people are found to reside.

By tracing who engages with the river, we can also see different patterns of behaviour and risk (Figure 2.8). During the dry season, it is in the socially deprived areas of the Tongi slum and Konabari where women are found to be at the river undertaking their daily chores. Men are found at the river during the Bishwa Ijtema to join the religious festival and face similar health risks. During the wet season, the assumption that river quality improves with higher flows may contribute to boys and girls swimming in the water in the Tongi slum, though scientific data show the pollution risks remain. Men also engage with the river, beyond the daily boating duties, across all the zones, though most modestly in Mausaid. The few observations of women, girls, men, or boys in the rivers near Mausaid partly both reflects better water and sanitation facilities in the home but also higher development and social indicators. As wealth and welfare increases people choose to live in cleaner river environments and they do not interact with the dirty rivers. This story is the same in Dhaka as it was before in Paris, London, or Washington, D.C.

Government, industry, and donors are aware of the growing water insecurity inequalities and collectively have plans to invest over USD 20 billion in water treatment infrastructure in the coming decades (Byron and Yousuf, 2022). This is to be welcomed but with certain caveats. First, the time frames for infrastructure investments are generational demanding early action for the most vulnerable. Second, regulatory non-compliance of environmental pollution is normalised and unlikely to change without unprecedented political or legal reform. Third, the ready-made garment industry looms large as a key, but not unique, actor in improving environmental, social, and economic outcomes.

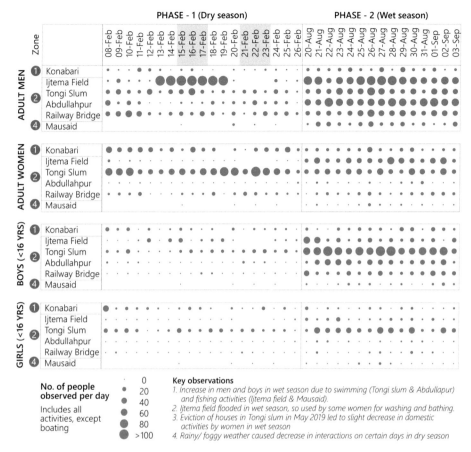

Figure 2.8   Observations of river use disaggregated by gender and age. Reprinted from Hoque et al. (2021) under the terms of the CC BY 4.0 license.

Social inequalities in Dhaka appear overwhelming in the dense fabric of a city that continuously grows and expands. Addressing the inequalities in Abdullapur, and other areas, needs to be an investment priority. Sharing water and toilets in such dense settlements increases public health risks in addition to reducing human dignity and welfare. Development actors have the means and the money to make significant improvements to bring safe water and sanitation services to these people. This will not immediately address the health of rivers but will make a significant contribution to improving water security in a few years, not a few decades. These areas also have high concentrations of garment workers living in extremely harsh living conditions.

Second, tackling regulatory non-compliance to environmental regulation will be a major challenge, partly because many politicians and factory owners are the same people. Unravelling this political conundrum is unlikely to be quick or easy.

While the ready-made garment industry reports billions of dollars of revenue, the margins are thin. Water treatment costs money and workers' wages are already low and under pressure to increase. Bangladesh performs well globally in the sector due to high-quality and low-cost labour. The economic calculus is challenging; as external regulatory costs are increased by Europe and other global markets, the industry has to manoeuvre agilely, which may lead to selling to less regulated markets in Asia. With an estimated 6,000 factories in Dhaka, the global excellence of the environmental standards of a handful of factories belies the significant damage caused by the majority. If economic history is a guide to the future, Bangladesh will gradually exit the market in decades to come as China and other countries have done.

However, the process of industrial change is highly uncertain and the well intentioned efforts to relocate the tannery industry was not executed well, for a range of reasons, with river health damage relocated from the Buriganga River to the Dhaleshwari River. Public frustration with progress has grown leading to popular support for the legal rights of rivers. It remains unclear if legal precedent can triumph where politics and industry has failed. Global experience of the rights of rivers is emerging and mixed. India is probably the most relevant comparison to Bangladesh with politicians deciding to revoke legal instruments for rights of rivers on the Ganges River in favour of a massive programme of infrastructure investment for water treatment facilities (Mishra and Upadhyay, 2021).

Beyond the feasibility of legal action improving river health, a proximate victim of action can be the communities living illegally on the riverbanks. Forced removals have occurred on the Tongi slum in Dhaka since the passage of the law. Where these people now live is unclear though their livelihoods have been radically disrupted with limited material change to the health of the river. Since the horrific Rana Plaza disaster a decade ago, global brands working in the industry now take a more direct and proactive role in factory conditions and worker welfare. Understanding and engagement with river health has increased though operates at the factory level, or cluster of factories, rather than the river system. Understandably, there are many other sources of river pollution with the vast quantities of untreated sewage an obvious priority.

Ultimately, the health of Dhaka's rivers is a government responsibility. Europe and North America massively polluted their urban rivers in their own race for economic improvement. These previously damaged rivers have now recovered where government regulation and enforcement remain strong and consistent. The Government of Bangladesh has the means and the authority to twin-track the longer-term improved health of the rivers and the short-term improved water security in the most polluted locations. The government's Delta Plan 2100 provides a long-term planning and financial framework for action modelling uncertainties and interactions in monsoon patterns, demographic variability, economic growth, and social welfare (General Economics Division, 2018). Our river observations show the granular and daily reality for people living in extreme hardship today. The moral imperative to act today for the most water insecure people is perfectly compatible with an environmental obligation to progressively improve river health over the next decade.

3

Chronicles from the Coast

Public and Private Responses to Water Risks in Khulna

3.1 Introduction

About 300 km south of Dhaka, away from the chaotic city life, is Khulna – one of the 19 districts officially demarcated as 'coastal' owing to the influence by tidal processes from the Bay of Bengal. While most of Dhaka's residents are detached from the polluted rivers in their day-to-day urban lives, rural lives and livelihoods in the coastal region are intrinsically linked to the hundreds of tidal rivers and creeks that meander through a fragmented landscape of embanked islands called 'polders'. Constructed during the 1960–1970s, the network of embankments and sluice gates were designed to control flow of saline water into the low-lying agricultural lands or '*beels*'. By increasing food security, improving road communication, and offering protection from storm surges, the polders provided a sense of permanence, encouraging growth of population and settlements. Today, more than 8 million people live across the 139 polders covering a land area of 12,000 km^2 (BWDB, 2013).

These densely populated deltaic floodplains are, however, prone to multiple water risks – from rapid-onset extreme events like cyclones and storm surges to chronic drinking water scarcity due to salinity (Hoque and Shamsudduha, 2024). The water crisis peaks during the driest months of March to May, when months of no rainfall limit recharge of ponds, with decreased upstream flows from the Himalayas causing saline water from the Bay of Bengal to fill the tidal rivers. While groundwater is easily available at shallow depths, salinity often exceeds the drinking water threshold, especially in the southernmost areas close to the Sundarbans mangrove forest. Freshwater may be available at deeper depths, though there are large variations in aquifer availability within short distances. Rainwater offers temporary relief during monsoon, with 2,000 mm of rain occurring within a short period of four and a half months between June and mid October (Shahid, 2011). In a landscape surrounded by water all year round, the scarcity of drinking water is indeed ironic.

The public sector response to these environmental risks has been to improve water supply through infrastructure investments. Since the 1990s, the government and international donors have poured in huge sums of money to build a diverse range of water supply technologies based on the local hydrogeological conditions. In line with the global policy discourse, this infrastructure-led agenda has indeed been successful in increasing access to improved water sources throughout rural Bangladesh, from 65 per cent in 1990 to 97 per cent in 2015 (GED, 2015). Much of this progress can be attributed to the remarkable adoption of tube wells – a transition that was initiated by the public sector but later fuelled by private self-supply investments by rural households. Yet, policy and planning documents consider publicly financed and community-managed water systems as the dominant institutional model for rural water service provision, with privately owned and managed sources not being formally recognised or monitored. Other forms of private sector engagement, such as the growing market of small water enterprises, also remain outside the scope of national accounting. In the absence of any regulatory oversight, this reallocates the responsibilities of financial and operational risks to individual households, schools and healthcare facilities, jeopardising safely managed drinking water services for the most vulnerable.

In this chapter, we draw on the daily water diaries of rural households in Khulna district to present how water source choices vary by local hydrogeology, season, and socio-economic status. We ground these behavioural dynamics within the uncoordinated landscape of public and private investments in water supply infrastructure, with uncertain water quality risks from salinity, arsenic, manganese, and *E. coli*. We argue that the existing institutional arrangement for water supply 'provision' in rural households and schools needs to be reformed to address the inequalities in access, environmental, financial, and operational 'risks' to safe and reliable services for all. In doing so, we share our experiences of designing, piloting, and scaling a results-based professional water service delivery model for schools in Khulna district. This story of change showcases how high-level government commitment, catalysed by donor funds and science-practitioner partnerships, can trigger in shifts in practice.

3.2 Salinity Risks and Investments in Water Supply Infrastructure

It was mid February 2017, just days after the onset of *Falgun* (spring) season. As we drove down the embankment road along Polder 29, the greyish sediment laden river flowed along the outer side, while vast stretches of paddy fields covered the landward side, often interspersed with settlements. The quintessential Bangladeshi village home comprises mud or tin-walled houses with thatched roofs surrounding a courtyard, shaded by fruit trees. A cluster of houses is likely to be inhabited by

Figure 3.1   Sisters-in-law busy with chores on a typical afternoon in Polder 29.

family members of the same paternal lineage, though each house may belong to a separate household. Cattle munching on haystacks, with cow dung sticks being sun-dried to be used as cooking fuel, are symbols of prosperity. Small ponds can be found among these clusters, with the water being used for washing, bathing and sometimes cooking. Somewhere in the courtyard, there is likely to be a hand pumped tube well – the primary source of drinking water for millions of rural Bangladeshis (Figure 3.1).

Bangladesh's tube well story started in the 1980s – the International Drinking Water and Sanitation Decade. The global political framing of risk at that time revolved around diarrhoeal diseases from drinking microbiologically unsafe water (Fischer, 2019), particularly among infants and children under five, prompting increasing allocation of international aid to developing countries in Africa and Asia to improve access to safe drinking water. The quest for the appropriate low-cost technologies prompted the design and field testing of different handpump models based on the hydrogeology and cultural preference of the place. For Bangladesh, where the water table is very high, UNICEF's No. 6 handpump met all the desired criteria of low installation cost, easy operation and maintenance, and durability (Black, 1990). It is a simple robust suction pump that can be manually drilled through the sand and silt using the sludging method and can be easily repaired as all moving parts are above the ground.

The installation of tube wells was initially led by the Department of Public Health and Engineering (DPHE), the national lead agency mandated for provision of rural water supply, with technical support from UNICEF. However, once the government released its control over local supply chains, the private sector took over, resulting in the growth of domestic production and maintenance capability (Fischer et al., 2020). The easy availability of spare parts and masons, coupled with increased demand for private water sources within one's premises, fuelled the growth of self-supply through shallow tube wells (less than150 m below ground level). Today, it costs around USD 125 to install a shallow tube well via private drillers and USD 275 when installed by DPHE. In comparison, deep tube wells cost more than USD 1,100, thus, their installation is either financed by DPHE or well off individuals. With 1.74 million publicly funded tube wells in 2019 (DPHE, 2019) and an estimated 18 million private tube wells (Fischer et al., 2020), Bangladesh's tube well story is unprecedented.

Tube wells, however, have been less successful in improving drinking water access for the coastal inhabitants. While there is a general trend of increasing salinity towards the Bay of Bengal, the coastal aquifer system is very complex characterised by high degree of spatial and vertical heterogeneity. In general, there are three layers based of depth and the nature and geological age of the sediments – (1) the upper shallow or first aquifer, formed about a 100 years ago and extending down to 50–100 m below the surface; (2) the second or main aquifer, dated as about 3,000 years old, and occurring up to a depth of 200–300 m; and (3) the deep aquifer, estimated to be about 20,000 years old and investigated up to a depth of 350 m (Aggarwal et al., 2000, BWDB-UNDP, 1982). Salinity in the shallow aquifers is extremely variable both spatially and seasonally, often exceeding the official permissible threshold level of 1,000 ppm of total dissolved solids (or electrical conductivity of 1,500 μS/cm). Vertical infiltration of rainwater during monsoon recharges the aquifers, with isolated freshwater pockets from recent precipitation indicating that the aquifers are not hydraulically connected. The deeper aquifers exhibit a more uniform distribution of salinity and continuity on a regional basis, though much of it remains understudied (Akhter et al., 2023, Ahmed, 2011).

In absence of systematic hydrogeological records, inferences about the aquifer system can be drawn from the depths and salinity of existing tube wells. Our water infrastructure mapping in Polder 29, which included all 2,805 public and private tube wells in the southern half (Figure 3.2) and a selected sample of 354 tube wells in the northern half, showed progressive increase in salinity in the shallow aquifer along a north-south transect with a corresponding decrease in the thickness of the deep aquifer. As such, 74 per cent of the tube wells in the northern half were used for drinking compared to only 38 per cent in the southern half, though 55 per cent of all drinking water tube wells exceeded the recommended salinity threshold of 1,000 ppm. The impacts of salinity became more visible as we continued our journey towards the southern end of the polder. The green paddy fields were

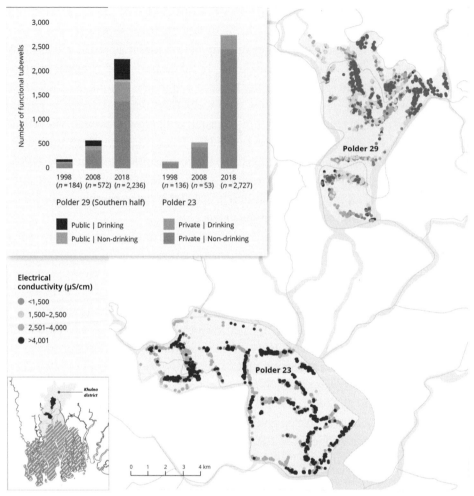

Figure 3.2   Location and water salinity of tube wells mapped in Polder 29 and Polder 23 of Khulna district.

replaced by fallow land with cracks appearing on the topsoil, while a few cattle grazed on the remaining hay. Most houses in the area had provisions for rainwater harvesting, though only better-off ones could afford large plastic storage tanks. It has been months since the last rains, and with the dry season about to intensify in March–April, most households, except for those with tanks, would soon run out of their stored rainwater.

Community water sources in these high salinity areas include pond sand filters and small piped schemes, mostly installed by local NGOs through donor-funded projects. The first pond sand filters were installed by DPHE in Dacope upazila (sub-district) in 1984, with subsequent modifications in design. A pond sand filter comprises a filtration chamber with a layer of sand and brick chips, which

Figure 3.3   Women using *kolshis* and plastic bottles to collect water from a pond sand filter (Photo credit: Lutfor Rahman).

treats surface water from rain-fed ponds that can then be abstracted by a hand-pump mounted on a raised platform (Figure 3.3). As of 2019, there are about 3,500 pond sand filters in the three southwestern coastal districts of Satkhira, Khulna and Bagerhat, though data on functionality is unavailable (DPHE, 2019). For rural piped schemes, donor investments have increased since the 2010s, though recent data from UNICEF/MICS (2019) shows that only 5.5 per cent, 5.3 per cent and 0.7 per cent of the surveyed households in Khulna, Bagerhat, and Satkhira districts respectively use piped water as their main source. We mapped a total of 49 piped schemes in Khulna district in 2022, of which three are in Polder 29 serving between 150 and 300 households within a 1–2 km of the source borehole.

With improved living standards, there have been growing demands for vended water. During the dry season, locally made three-wheeled motorised vehicles called *nossimons* or pedalled vans can be frequently sighted carrying stacks of blue or white coloured water containers. One of the most popular sources for vendors in Polder 29 is a motorised tube well, located in a mosque just beside the river. Owing to its location, water from this tube well is also transported via trawlers to other polders further south (Figure 3.4). To interview the vendor and track his route, we took a ride on the trawler with about a hundred 30-litre containers and a few 200-litre drums of water. After an hour of sailing, as we approached Polder 22, we could see men, women and children lined up on the embankment with *kolshi* (aluminium pitcher), buckets, and bottles. As soon as the trawler docked on the jetty, they swarmed in to fill up their containers before the water ran out.

Figure 3.4   Water from a deep tube well in Polder 29 being transported in 30-litre containers via a trawler to be sold to villages 6–8 km further south (Photo credit: Lutfor Rahman).

Tube wells are not the only sources of vended water. Since the late 2010s, there has been a boom in investments in reverse osmosis-based desalination technologies by donor organisation and grassroots entrepreneurs. We mapped 63 desalination plants in Paikgachha upazila (further south of Polder 29), of which 93 per cent were privately financed by local residents who saw a business opportunity while also addressing water scarcity in their communities (Hoque, 2023). Several factors converged to create a business opportunity for this niche innovation to flourish, including economic growth resulting from shift to export-oriented aquaculture, expansion of rural electrification to remote areas, and availability of cheaper reverse osmosis technology imported from China.

Analysis of the water supply technologies and capital expenditures in both Polders 29 and 23 show a clear trend of increasing private investments with increasing salinity from north to south. Between 2010 and 2020, USD 385,000[1] and USD 200,000 were invested by the government and donors in Polder 29 and Polder

---

[1] Exchange rates for BDT to USD are based on the time of data collection and range from USD 1 = BDT 80–100 in this chapter.

23, respectively, compared to USD 252,000 and USD 410,000 being invested by households and local entrepreneurs (Hoque, 2023). Many a times we have heard government officials and residents saying, 'We suffer a lot for water, because tube well is not 'successful' in our area', resonating the institutional and cultural mindset that tube wells are a symbol of water security. Although piped schemes are the future of rural water supply, whether through public taps or household connections, the inherent inertia to invest public funds in non-tube well technologies can be partly attributed to bureaucratic hurdles. On one occasion, a local engineer of DPHE mentioned, 'We got an allocation for 26 tube wells per union. If we drill and cannot find any aquifer, we cannot reimburse our contractors. So, we have to install these in areas known to have good groundwater. We cannot pool these funds and install a piped system unless directed by specific projects.'

These variations in water infrastructure type, as dictated by the local hydrogeology, is also reflected in water services for rural schools in Khulna district (see Section A.5 of Appendix for methodological details). The institutional and financing arrangement is more streamlined for government primary schools than for secondary or non-government schools. For government primary schools, capital investment for water and sanitation facilities is financed through the Primary Education Development Programme – a multi-donor and government funded project supporting the primary education sector since the early 2000s, and the installation process is led by DPHE (Fischer et al., 2021). Secondary schools often have integrated ownership, meaning that the government pays for staff salaries while other costs need to be covered through tuition fees. Thus, waterpoints in secondary schools, as well as in private and NGO schools, are installed through school funds, donor-funded projects, and funds from union council or local Minister of Parliament, and in some cases by DPHE, if there is no boundary wall restricting access to the community.

Schools in the northern part of Khulna district often have multiple tube wells, upgraded with submersible pumps and internal plumbing. Those in high salinity areas have rainwater harvesting, but the availability is restricted by tank size and roof catchment area. In Dacope, one of the nine upazilas, many schools have high storage capacities (more than 10,000 litres) owing to multiple donor-funded projects operational in that area for decades. Other technologies, such as reverse osmosis plants, small, piped schemes and pond sand filters, are available in small numbers. For the 15 per cent schools that do not have any drinking water source, teachers' pay out of pocket to purchase vended water, and students may bring water from home. In many cases, children end up using unsafe sources like shallow tube wells or fetch it from nearby households where available.

With such diverse range of technologies for households and schools, it is evident that significant capital investments have been made by the government, donors, and the private sector over the past few decades. While this infrastructure-led response has surely met the agenda for increasing 'access' to improved sources', to what extent these water services are safe, reliable, affordable and equitable remains unclear. We now explore the social, spatial, and seasonal variations in household source choices and outcomes to unpack who bears the greatest risks, where, and when. These behavioural dynamics can highlight the gaps in existing institutional design and the needs for redistributing the risk responsibilities among the different public and private actors.

3.3 Spatial and Seasonal Distribution of Water Risks among Households

As discussed earlier, drinking water is not a concern for those living in low salinity areas, such as the northern half of Polder 29, where all households rely on deep tube wells all year round (Figure 3.5). Most of these deep tube wells are public, with 30 per cent being owned by an individual or group of households. In areas where groundwater is too saline to drink, such as the southern part of Polder

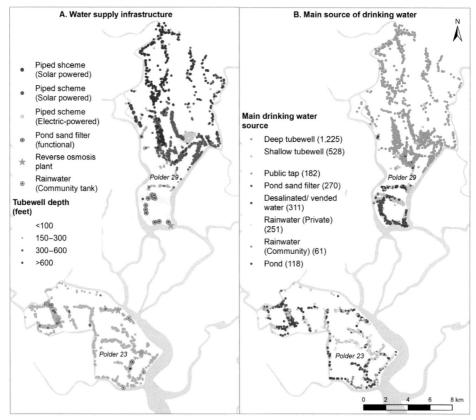

Figure 3.5   (a) Water supply infrastructure and (b) main sources of drinking water in Polder 29 and Polder 23.

29 and our other study site Polder 23, households use multiple sources through-out the year with varied implications for quality, costs, and distance (Figure 3.5). Though widely acknowledged, such variations in water sources are never cap-tured in aggregate statistics used for decision-making. Large-scale surveys, includ-ing the 10-yearly national census, the Demographic and Health Surveys and the Multiple Cluster Indicator Survey (MICS), which are funded by the Government of Bangladesh, United States Agency for International Development, and United Nations Children's Fund (UNICEF), capture access to the drinking water facilities and service levels, in terms of the main source, the collection time, water treat-ment methods, and recently, the reliability of the service. This results in a flawed portrayal of progress, suggesting that 98 per cent of rural Bangladeshis use a safe water source (GED, 2015).

Our water diary study, which included the water-stressed villages in south of Polder 29, revealed the day-to-day changes in water sources and costs. We identified significant spatial clustering in households' water source choices (refer to Section A.3 of Appendix for data analysis methods), indicating proximity to waterpoints and rainfall as the key behavioural drivers (Figure 3.6). Cluster 1, comprising 27 per cent of households, predominantly used vended water for drink-ing and shallow tube wells for other uses. In contrast, those in Cluster 3 (28 per cent households) reported using shallow tube well for drinking and pond water for other uses. This reflects the variation in groundwater salinity within short dis-tances. While water from shallow tube wells in this region is generally too saline to drink, as is the case for Cluster 1, one private shallow tube well (locally known as *Kalar kol*) located about 1 km north of Cluster 3 had low salinity, making it a lifeline for neighbouring villages. The greatest share of households (40 per cent) belonged to Cluster 2, using pond sand filters as the main source for both drinking and non-drinking uses. The remaining 4 per cent households in Cluster 4 were located further north, and thus, unlike others had access to deep tube wells for drinking. Use of rainwater peaked between July and September for all clusters, particularly Cluster 3, who could avoid walking to *Kalar kol* for a few months. Owing to the curvature of the river, the area of the polder inhabited by Cluster 3 households is prone to riverbank erosion and storm surge inundation. With many households having lost their land and living on the embankment, the area had lim-ited scope for installing water supply infrastructure.

The prevalence of chemical and faecal contamination varied across these differ-ent types of sources (Figure 3.7). The water quality in pond sand filters, the most widely used technology in the southern part of Polder 29 (refer to Cluster 2), var-ies substantially from one source to another depending on maintenance of source pond and the filter media. In general, they are prone to faecal contamination, par-ticularly in the rainy season owing to seepage from nearby latrines or livestock.

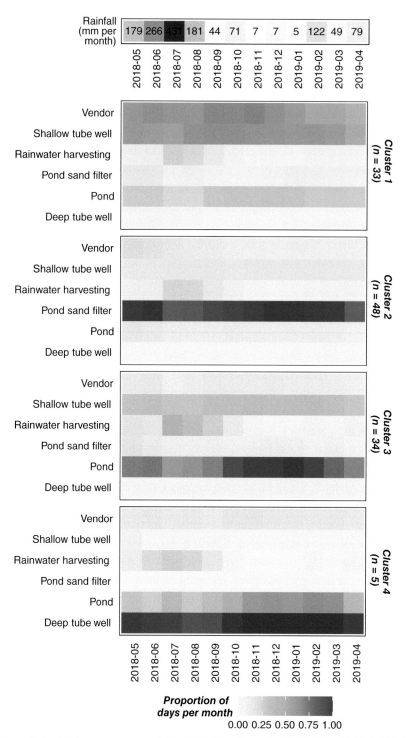

Figure 3.6 Water sources used by 120 diary households during 2018–2019 in relation to rainfall.

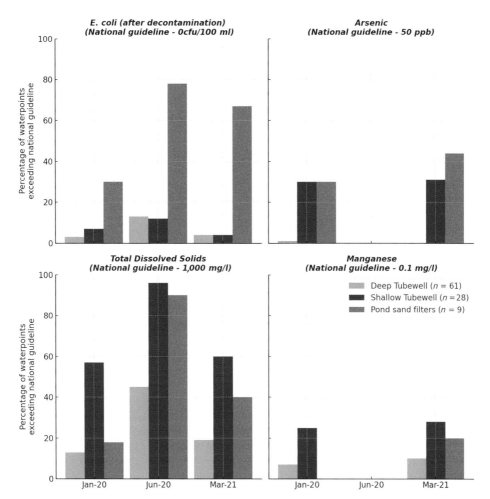

Figure 3.7   Seasonal variations in chemical and faecal contamination across 97 waterpoints in Polder 29.

Inundation by saline water from storm surges poses another challenge, as exhibited by June 2020 samples collected after cyclone Amphan in the previous month. Vended water from deep tube wells and reverse osmosis plants are generally free from contamination, though many prefer the former owing to better taste. As one participant explained, 'The vended water [from Sarappur tube well] is as natural as coconut water, but this [desalinated] water is treated with chemicals.'

These variabilities in water quality by source and time of the year, however, are not reflected in national statistics. During the MDG era, assessment of safety was simply based on whether a source was 'improved' or 'unimproved', thus, making tube wells safe by default. Starting from 2012, DPHE's Annual Waterpoint Status reports show upazila-wise 'coverage by safe waterpoint', where 'safe' is defined as having arsenic below the national guideline of 50 µg/l. These statistics are based on test results for public waterpoints at the time of installation, as there are no provisions for routine quality checking even on a sample basis. Indicators like salinity and faecal contamination are neither measured nor reported. Donor-funded or privately installed sources, which serve the bulk of the population in water stressed areas, are completely unaccounted for. The MICS 2019 was the first nationally representative survey to test for *E. coli* at the point of use. Results indicate that when arsenic and faecal contamination are considered, only 43 per cent of households use a safe water source on premises, down from the widely reported statistic of 98 per cent having access to improved sources (BBS/UNICEF, 2021).

Indicators for affordability, such as household water expenditures, are not available either, though this is a concern for all countries and contexts without piped water connections. Cost is the main deterrent for using better quality vended water, ranging from USD 0.24 to USD 0.35 per 30-litre container (USD 8.0–11.7 per m^3) (Hoque, 2023). Of the 120 diary households, 23 per cent did not purchase vended water at all, 44 per cent purchased water regularly throughout the year, and 33 per cent restricted their consumption to the dry season or as and when needed. With high usage of vended water, households in Cluster 1 had a median annual expenditure of USD 56, which was four times that of Clusters 2 and 3 with USD 9 and USD 14 respectively (Figure 3.8). Such differences were not observable for food expenditures.

However, several idiosyncratic factors influenced household water sources choices and expenditures as well. One of the households in Cluster 1, who recorded high usage of vended water, described,

We buy our drinking water from Sarappur tube well, about 3 containers a week. For cooking, I normally get it from Zia's [pond sand] filter, but in the rainy season when the roads become muddy, I use the stored rainwater. My daughter and granddaughter have been staying with us for more than a year now. Zia is my daughter's husband. They got married almost 10 years ago, but she had difficulty in conceiving. She underwent several treatments and finally when she got pregnant, doctors told her to be in bed rest. So she moved back in with us. After the birth of our granddaughter, we had several parties and relatives visiting us. Our water expenses increased during that time. Our granddaughter is very precious. We in fact bathe the child using water from the new desalination plant.

Habits and individual preferences thus drive inter- and intra-household differences in choice of sources, particularly for rainwater and desalinated water. Rainwater harvesting was widely practised; while most households used 50-litre earthen pots (locally known as *motkas*), about 14 per cent had 200-litre plastic drums and 3 per cent had large 1,000–2,000-litre storage tanks. Whether the rainwater was used for both drinking and cooking or for cooking only varied depended on how well individuals tolerated it. While some cited it as their most preferred source, others

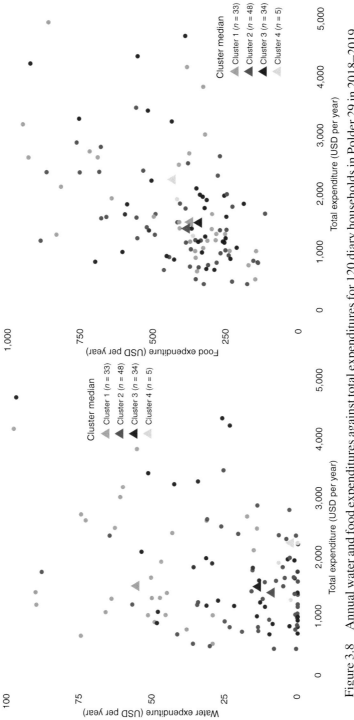

Figure 3.8   Annual water and food expenditures against total expenditures for 120 diary households in Polder 29 in 2018–2019.

complained about coughs, bloating or diarrhoea resulting from drinking rainwater. One household in Cluster 3 noted,

We normally drink rainwater in the monsoon and go to Kalar kol [Shallow tube well] during the dry season. But my son, who currently studies in Gopalganj [nearby district], cannot drink the water from any of these sources. So whenever he comes to visit, we buy water from Sarappur mosque [Deep tube well].

3.4 Institutions for Managing Operational Risks in Communities and Schools

Despite continued public and private investments in water supply infrastructure, households face unequal water risks, in terms of safety, costs, and physical burden of water collection, and these risks are exacerbated when the infrastructure becomes non-functional due to seasonal unavailability of water, source contamination, or technical malfunctions. The existing institutional arrangement dictates that, for publicly or donor-funded water infrastructure, users should undertake all operational and maintenance activities with monetary contributions from regular tariffs or ad hoc payments (Hope et al., 2021a). This community-based management model emerged in the 1980s as an alternative to the 'supply-driven' models of the post-colonial states that struggled to extend basic services to the expanding population. Community management was championed as a 'demand-driven' approach that decentralised responsibilities to local people by encouraging active involvement of users in construction and maintenance. Community water management was an expedient product of its time – a pragmatic response in many countries emerging from colonial rule in the 1960s with limited administrative capacity or financial resources.

The institutional culture of the community-managed model is one of 'egalitarianism', whereby users share the environmental, operational, and financial risks equally (Koehler et al., 2018). In Bangladesh, community management has been widely successful in case of tube wells owing to the simplicity of the technology and the easy availability of mechanics and spare parts. The handpump technology has a lifespan of 10–15 years. For high usage tube wells, fast-wearing parts such as washers, check valve, nuts and baseplate need to be replaced every few months. Given the low-price tag, the bulk of such costs are often borne by a local elite, with small contributions from users. This is in contrast with countries in Sub-Saharan Africa, where lack of local markets makes it very difficult to source spare parts, resulting in longer downtimes. We return to this later in the next chapter on rural Kenya.

Operational and financial risks are much higher for alternative technologies like pond sand filters and piped schemes, which in turn are often located in areas with high environmental risks from salinity. Pond sand filters require regular maintenance, which involves cleaning and replacing the sand beds and protect-ing the source pond from contamination. While pond sand filters do not have any fixed tariffs, users make small contributions for purchasing sand, replac-ing tube well parts, and associated labour costs when needed. However, not all users contribute which often creates resentments among those who pay. Monthly user payments and expenditures from nine pond sand filters during September 2019 and August 2020 averaged at BDT 500 (USD 6) (refer to Section A.4 in Appendix A1 for data collection methods). However, after cyclone Amphan in May 2020, one of the pond sand filters incurred a large expense of BDT 11,000 (USD 130) to hire machineries and labour to clear out tree branches and dead fish from the pond. Such large and unexpected expenses are often borne by a few relatively well-off households if there is good leadership from an active manage-ment committee.

Politics and local power dynamics can also exacerbate operational and financial risks, as seen in case of a solar energy operated piped scheme in Polder 29. This donor-funded scheme was executed by the Union Chairman who claimed it as a flagship project during his time in office and promised to serve water free of cost to win votes. While the scheme ran smoothly initially, it broke down after four years owing to a minor technical fault. Though the repair cost was trivial, the sys-tem remained non-functional for years owing to a lack of community ownership, which was further instigated by the new Union Chairman who blamed it on poor construction. During this time, users living close to the pumphouse reverted to a public tube well while those further away walked long distances or paid vendors to deliver water. Thus, when responsibilities for management are not clearly defined or borne by designated entities, the risks are passed on to users.

For schools, the responsibilities of operation and maintenance are borne by individual school administrators using funds from the annual school budget or the routine facilities maintenance fund allocated every three years. The institu-tional culture of these 'bureaucratic' management model is 'hierarchical, with rational procedures and decision-making tools in place to avoid extreme uncer-tainties (Koehler et al., 2018). For waterpoint management, this translates to rou-tinised operation and maintenance activities using dedicated funds, though this is not always the case in practice. With limited budgets of BDT 50,000–70,000 (USD 500–700) per year, government primary schools struggle to undertake any major repair or rehabilitation work, other than replacement of tube well parts as needed, and cleaning of rainwater tanks and catchments prior to onset of mon-soon. On average, small tube wells parts such as washers, buckets, nuts need to be replaced every four months, costing BDT 500 (USD 5) per repair, while large parts like tube well body or handle may need replacement once every four years, with costs of BDT 2,500 (USD 25) per repair (REACH, 2023a). In contrast, operating and maintaining a reverse osmosis plant, which involves frequent replacement of media and dosing media, may cost up to BDT 50,000 (USD 500) per year which is beyond what most schools can afford.

3.5 Reallocating Responsibilities to Professional Service Providers

The disparity in risks associated with water services for households and schools is profoundly demarcated by geographical and seasonal variations. This encompasses factors such as water availability, safety, and the costs of infrastructure installation and maintenance. Yet the institutional and financing model for bearing these risk responsibilities are similar, creating vast inequalities in services. While a deep tube well in a low salinity area can provide safe water all year round with negligible cost or maintenance responsibilities for users, those using a piped scheme have to pay for and maintain a relatively complex infrastructure for a limited quantity of drinking water only. In areas lacking suitable technologies, an 'individualist' risk management culture emerges, whereby informal water vendors earn a living by selling water to those who can afford to pay. If the risks can be pooled together at an appropriate scale (e.g. at district or upazila level), and the responsibilities are allocated to one entity (i.e. a private organisation contracted by the local government), there is opportunity to improve the safety, reliability, affordability, and equity of services.

Globally, especially within urban contexts, non-governmental entities have been incorporated into the water supply chain through diverse public–private collaborations or full-fledged privatisations. Professional maintenance service providers are emerging globally using results-based funding from donors and foundations in 16 countries in 2023 (McNicholl and Hope, 2024). An example is the Kenyan social enterprise FundiFix, responsible for timely repair and maintenance of handpumps and piped schemes in exclusive service areas in Kitui and Kwale counties (REACH, 2016). Developed by colleagues at the University of Oxford since the early 2010s, the experiences and lessons learnt from the FundiFix model facilitated the design and piloting of a professional service delivery model for coastal Bangladesh. While rural Bangladesh benefits from easy access to spare parts and skilled technicians for handpump repair, more intricate technologies like piped systems and reverse osmosis plants present challenges that require extensive technical and financial support. Another major risk, prevalent in both countries, is that of water safety, and in absence of monitoring, the risks remain uncertain.

This led to the conception of the SafePani model, advocating for a shift from the infrastructure-led approach of building access to a professional service delivery model (Hope et al., 2021a). The proposed change would entail DPHE to transition from direct service provision to monitoring and regulation. Concurrently, the responsibility of operation and maintenance could transition from local communities and schools to professional service providers operating in an exclusive service area under contractual agreements with the government. Monitoring and regulation need to be supported with better information systems on infrastructure coverage, functionality, water quality, and costs. At the same time, the sector financing can include results-based contracts where government and donor funds pay after delivery of agreed service targets, such as reliability (uptime) and water safety. In late 2021, we initiated a two-year pilot programme to demonstrate the SafePani model in Khulna district. Through funding from the REACH Programme, the Bangladeshi non-profit organisation HYSAWA was contracted to deliver professional maintenance services to all government schools and healthcare facilities in eight selected unions (REACH, 2023b). We opted to showcase the SafePani model within government institutions due to the uniformity in their governance structures, which in turn, facilitated the scalability of the model.

The SafePani pilot phase had three overarching objectives.

First, we aimed to demonstrate the operation of the model in practice. This involved preventative maintenance of water points and prompt repair services upon problem identification, monitoring water safety through sanitary inspections and laboratory tests of microbial and chemical parameters and taking remedial actions like shock chlorination upon detection of faecal contamination. Where needed, existing water supply systems were first rehabilitated to acceptable standards, for example, by reconstructing tube well platforms or resurfacing rainwater catchments.

Second, we engaged with relevant government agencies from central to local levels to introduce the idea of professional service delivery. Since the inception of the pilot phase, a national steering committee and district working group were formed, with quarterly meetings to report progress and plans. If any actions were necessary but beyond the scope of SafePani, for instance, installation of new water supply infrastructure in institutions lacking any drinking water sources, they were reported to relevant government agencies.

Third, we aimed to build technical and management capacity for professional water service delivery at local level. In doing so, we formed a local team comprising engineers, managers, community mobilisers, and water quality technicians to lead the day-to-day tasks of service delivery. We also established a purpose-built laboratory in the SafePani Khulna office to test for *E. coli*, while responsibilities for chemical tests were borne by DPHE district laboratory.

The pilot phase demonstrated that professional service providers can ensure water safety and reliability, with a cost of less than USD 1 per student per year (REACH, 2023a). It garnered government interest to upscale the model to 1,200 schools and healthcare facilities in Khulna district through results-based funding. In 2024, the government formally committed to fund 45 per cent of costs for the district scale-up till 2030, the rest coming from donors managed by the Uptime Catalyst Facility, a UK registered charity. The SafePani model is flexible in design with its current focus on public facilities able to expand to local villages as results and resources provide the evidence for a universal approach over time.

Professional service delivery will be of significance for operation and maintenance of rural piped schemes, which are particularly critical for areas without good groundwater. Our audit in 2022 found that 27 of the 49 rural piped schemes in Khulna district were non-functional, equating to USD 1.2 million in wasted capital funds in addition to the daily coping costs borne by unserved households. It is logical and politically expedient to introduce professional service delivery before new water supply construction begins to provide an accountable and mutually acceptable approach to increase the returns on investment.

3.6 Conclusion

Living in coastal Bangladesh is a good working definition of being water insecure. Cyclonic storms and storm surges of various intensities batter the flat floodplains pre- or post-monsoon. Salinity in groundwater and tidal rivers constrains drinking water supplies, particularly in the dry summer months. The region's high vulnerability to climate change has increasingly attracted donor and NGO activities, with a cycle of infrastructure interventions for flood protection embankments and drinking water technologies.

The diaries offer daily and seasonal insights into the risks and responses to increase water security. First, we can see the distributional inequalities and individual choices by households to navigate their water insecurity. Second, there is a growing understanding of how water safety is a foundation to achieving water security from both natural contamination and bacteriological risks. Third, there is optimism and evidence that progress can be made by working closely with government partners to deliver safe drinking water for schools and health clinics.

Given a choice, most rural people prefer not to pay for water. If there is a free or low-cost alternative, regardless of the real or perceived quality, this tends to drive daily decision-making. Summer rains provide a good source of high-quality water collected at home, though contamination is likely to be high, without good hygiene practices. Drinking water payments are commonly driven by scarcity and aesthetics (taste, appearance, smell) not concerns of microbial or chemical contamination. Desalination or vendors provide a lifeline to the most vulnerable communities, though at a cost. Few households choose to regularly pay for vended services, even though it is a fraction of other costs, such as food. The argument that water is unaffordable has to be balanced against these behavioural choices and should avoid the assumption that people won't pay because they can't pay. Cultural practices matter. We see the health of a baby or the arrival of a close family member shifts water payment choices.

Unlike food expenditure, drinking water payments rise and fall during the year. Again, one has to be careful with assumptions and causality. Most households in water scarce areas have no regular access to a good service all year. While infrastructure access is high, water quality and reliability are often low. Why would anyone pay from limited funds for a poor service? Without providing an inclusive and safely managed drinking water service, it is premature to assume that demand is low. However, the legacy of policy driving access means households have made considerable investments in local technology, the shallow tube well. Displacing this investment will be difficult without a superior and low-cost alternative.

This brings us to our second reflection on water quality. Water insecurity in coastal Bangladesh is heavily influenced by water safety concerns. As noted, past policy failed to take water quality adequately into account in agreement with global development goals and donor funding. The bleak recalculation of Bangladesh's progress is halved once water quality is factored into the arithmetic. From nearly everyone with safe drinking water to less than half in the stroke of a pen. To the government's credit, they are making meaningful progress to track water quality (*E. coli* and arsenic) in nationally representative surveys to guide future policy and investments. Given the widespread occurrence of highly saline water with major health risks for many vulnerable groups, it would be relatively cheap and easy to track salinity in addition to the efforts being made on *E. coli* along with known risks from arsenic and growing concerns with manganese (BBS/UNICEF, 2021).

This leads to some positive progress in our third reflection. Working closely with national and international scientists in the REACH Programme, the government has successfully piloted a safely managed drinking water model for rural schools and health clinics. While the majority of investments have historically targeted drinking water in villages, there has been a blind spot with public facilities. As any epidemiologist will tell you: it is not how much good water you drink but how much bad water you drink. Safe water in the village and unsafe water in public facilities will generate limited education, health, or economic benefits. Public facilities provide a less political and more scalable approach to introduce safe water services than villages. In the latter, local management, including the associated revenue, makes changes slow and contentious. Even bad water makes money where few alternatives exist. Government control is limited in making changes at scale quickly in contrast to public facilities where consensus based on clear science and policy mandates is simpler.

The SafePani model offers an inclusive architecture where the major gains in improving water safety and reliability can be scaled out to neighbouring villages who can see the improvements being delivered. In parallel, if the district scale- up in Khulna is successful it could be applied to 65,000 public primary schools and 18 million pupils nationally. Such progress may take time and also provide the means to make strategic investments in piped water systems which would greatly simplify water safety in the future. Given Bangladesh's high population density, a networked system managed by a professional service provider could rapidly transform water security at scale quickly and relatively low cost. Neighbouring India provides a sense of how political momentum and investment can deliver results at scale quickly. However, the ultimate prize is sustaining the services for decades to come for which Bangladesh has carefully designed sustainability into the SafePani model.

4

Watering White Elephants

Rainfall Revenue Dynamics for Rural Water Services in Kitui

4.1 Introduction

Clear blue skies, reddish-brown soils, and bright yellow jerrycans – the three colours of rural Kitui that symbolise the water stories of these semi-arid landscapes. With extremely low and variable rainfall that disappears quickly upon touching the hot sandy ground, searching for and fetching water often dominates the daily lives of women and children in these hinterlands of Kenya. This is a story that spans the ages in rural Africa from the biblical reference to the 'drawers of water' to the more recent critique of 'watering white elephants' (Therkildsen, 1988). The former captures social inequalities and hardship that have barely changed in centuries despite Kenya's major investments in ports, railways, roads, and devolution in recent years, leaving the country close to an international debt default in early 2024. The latter is homage to Therkildsen's detailed critique of ineffective and uncoordinated donor investments in Tanzania in the 1980s, constructing rural water supply 'white elephants' that crumbled soon after the projects concluded, and the donors departed. As in Bangladesh, it is convenient to use the climate crisis as the primary villain in rural water insecurity, but the reality is that ineffective governance and weak service accountability provide important clues and responses to breaking a dismal cycle of wasted investments and underdevelopment.

Despite being traversed by the equator and close to the Indian Ocean, the unexpected dryness of East Africa's climate is intriguing. Recent evidence shows that such dry conditions can be attributed to the region's topography, whereby the network of valleys interspersing the 6,000 km long East African Rift System cause moisture laden air to be channelled at high speed towards Central Africa leaving an arid landscape in its path (Munday et al., 2023). Unlike the tropical monsoon in Bangladesh, spanning across four months between June and September, rainfall in Kenya is bimodal. There are two distinct rainy seasons – the 'long rains' from March to May, and the 'short rains' from October to December, which have

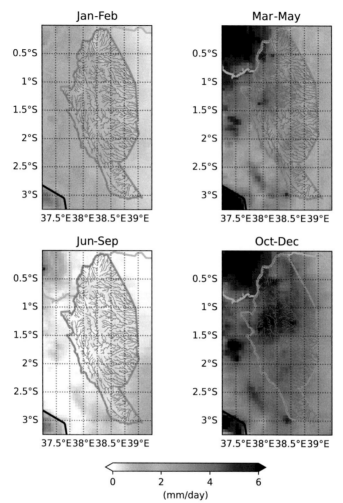

Figure 4.1 Spatial and seasonal variations in rainfall over Kitui county, illustrating the 'long rainy season' (March–May) and the 'short rainy season' (October–December) separated by a prolonged 'dry season' (June–September). Map drawn by Ellen Dyer using rainfall data from 2016 to 2022 available from the Climate Hazards Group InfraRed Precipitation with Station data (CHIRPS).

a combined annual average of 900 mm of rainfall (Figure 4.1). With most of the rivers being seasonal, digging out rainwater stored in dry riverbeds has traditionally been the only reliable water source for these sparsely populated rural populations.

Kitui, together with its neighbouring Machakos and Makueni counties, is largely inhabited by the Kamba people of the Bantu ethnic group. Though traditionally semi-nomadic herders, the Kamba were forced to shift to subsistence cropping along with limited livestock keeping during the British colonial era (Kisovi, 1992). Water for irrigation and livestock is key to survival, with a natural affinity for people to settle close to sandy riverbeds or valleys where rainwater can be harvested. Since the late twentieth century, there has been growing investments in earth dams, rock catchments and sub-surface sand dams in this region to augment rainwater storage, as part of an overall drive to improve productivity of the arid and semi-arid lands. Donor investments in groundwater technologies such as handpumps, boreholes with submersible pumps, and piped schemes with kiosks or standpipes also gained momentum since the turn of the millennium, as 'access to improved sources' gained centre stage in global water policy.

Yet as of 2019, 44 per cent of the 1.1 million residents of Kitui used surface water as their main source of drinking water, compared to the national average of 23 per cent (KNBS, 2019b). With such high reliance on unimproved sources, the global target of safe and affordable drinking water on premises by 2030 seems elusive in these contexts. Despite billions of dollars of investments in rural water infrastructure, why do one in three Kenyans still rely on surface water? How can sector financing and institutions be reformed to improve water security in these drylands with climate uncertainty? In this chapter, we explore the answer to these questions through analysis of the daily water sources and payments of households and schools in Kitui. We situate these behavioural dynamics within the changing political and institutional landscape of rural water service delivery from the colonial era through to the post devolution Kenya.

4.2 From Colonial Times to Harambee Habits

It was late March 2017, just days before President Kenyatta declared a national drought emergency. As we left the highlands of the bustling capital of Nairobi to pilot our water diary method, the signs of water stress became visually conspicuous. The unpaved dusty roads towards Mwingi-North were surrounded by uninhabited stretches of stunted thorn bushes, occasionally punctuated by Baobab trees. Looking down from a bridge, we saw men, women, and children digging scoop holes in the dry riverbed under the scorching sun. As they filled up their jerrycans pint by pint using funnels cut from plastic bottles, the donkeys waited to be loaded before making their long arduous walk back home. This is, in fact, a very common scene in these arid and semi-arid lands of rural Kenya, where sandy riverbeds act as a sub-surface reservoir for months after the last rainy season, preventing water losses through evaporation (Figure 4.2).

Built on impermeable layers of underlying bedrock, sub-surface dams can further augment the storage of rainwater within the highly porous sand collecting behind the dam wall. An audit of water infrastructure in Kitui county

Figure 4.2   People extracting water from scoopholes in a dry sandy riverbed in rural Kitui. The photo was taken in March 2017 just days before the Kenyan President declared a national drought emergency (Credit: Rob Hope).

identified close to 700 sub-surface sand dams (Nyaga, 2019), although the total numbers can be as high as 1,500, making Kitui a 'global leader in sand-storage technology, at least in terms of dam numbers' (Ertsen and Ngugi, 2021, p. 4). While most are unequipped, the more productive ones may be fitted with pumps, pipes, and tanks to supply the water off-site. Earthdams and rock catchments are popular rainwater harvesting infrastructure in Kitui. Built to contain surface runoff from valleys and slopes, they usually have water for only a few weeks after rainfall due to high evaporation. Construction of sand and earth dams are never intended for drinking water purpose only. Rather they are vital for small-scale irrigation and livestock, which are the main sources of income and sustenance.

The distribution of water resources and feasibility of different water supply infrastructure are determined by the topography, rainfall, and geology of the region. The Athi and the Tana are the only two perennial rivers in Kitui, with most of the seasonal rivers draining into the Tana River Basin. The topography of the landscape falls from a peak of 1,800 m in the western highlands to about 400 m towards the eastern plains, interspersed with deeply weathered metamorphosed rock outcrops. The lowlands which constitute two-thirds of the county receive less than 500 mm of rainfall a year – that is, a quarter of the tropical monsoon in Khulna. Rainfall amount and distribution is more reliable in the short rainy season from October to December than the March to May long rains' season. Knowledge of groundwater quality is scattered and limited to individual drilling records, without any systematic aquifer characterisation to aid the siting of boreholes. In general, salinity from naturally occurring chlorides, fluorides and nitrates is a major challenge, particularly in the low-lying plains with colluvial deposits and red soils (Wadira, 2020).

Prior to the colonial era, the Akamba people adapted to the region's limited and unreliable water resources through their flexible patterns of settlement and mobility, characterised by a mix of private and common property rights that supported integrated crop-livestock systems (Rocheleau et al., 1995). The arrival of the British colonial settlers disrupted the traditional land use and settlement patterns, marking a transition from 'people going to the water' to 'water going to the people' (Nyanchaga, 2016). Construction of the Ugandan railroad from the Mombasa port marked a pivotal shift in Kenya's water infrastructure development. The first piped systems were built around stations in the early twentieth century to meet the needs of the steam engines and the railway workers. Around the same time, the colonial government also launched extensive land seizure and enclosure operations aimed to limit mobility of livestock of the Akamba herders, whose 'primitive' cattle rearing and agricultural practices were framed to be the cause of the devastating soil erosion that plagued the native reserves (Rocheleau et al., 1995). The first rural piped schemes and boreholes were drilled by the British colonial administration in the mid twentieth century to facilitate commercial ranching within controlled grazing areas and intensive cultivation of cash crops. However, lack of planning and investment led to subsequent failure of boreholes and piped schemes, and reallocation of funds for constructing small-scale surface and sub-surface dams (Parker, 2020). As described by Munger (1950, p. 580), boreholes were 'entirely outside the native's traditional knowledge and psychologically … less desirable than dams'.

Since these early interventions by the colonial regime, the rural water sector in Kenya underwent several stages of institutional reforms in line with global water policy discourse and the evolving economic and political situation within the country. Water reforms in post-independence Kenya aimed to correct the colonial injustices by adopting African socialism as the development philosophy, with the concept of self-help, termed as 'Harambee' by the first President Jomo Kenyata, being the main vehicle for driving equity in rural areas (Nyanchaga, 2016). Self-help groups, which dominate the management of rural water systems till date, are generally groups of local residents with shared economic and social interests working together for their own betterment. Throughout the 1960s and early 1970s, self-help groups initiated many rural water projects through labour and financial contribution, while donor organisations like UNICEF and WHO supported the government to augment rural water infrastructure development

(Mumma, 2005). This philosophy also applied to schools, with the share of Harambee schools in Kenya increasing from 52 per cent in 1969 to 73 per cent in 1989 (Hope et al., 2021b).

4.3 Limited Reach of Policy Reforms in Rural Areas

The post-independence ideology of water as a 'social good' made it increasingly difficult to fund operations and maintenance costs in rural areas with the large distances between waterpoints. At the same time, the government's ability to finance rural water supplies was significantly constrained in the 1980s, as the country plunged into a debt crisis from accumulation of unpaid loans from the International Monetary Fund and the World Bank. This was further exacerbated by a severe drought that curtailed export revenues from tea and coffee. As with other newly independent states of the global south, the World Bank's Structural Adjustment Programme led to significant restructuring of Kenya's economic policies. In line with the overall market-oriented reforms aimed to reduce public spending, the rural water sector saw a push towards decentralisation of service delivery, higher degree of self-financing, and improved operational efficiency. Likewise, the government promoted a cost-sharing policy for schools, with teacher salaries and learning resources being funded by the government while responsibilities for infrastructure and recurrent expenditure being largely borne by communities (Ngware et al., 2007). As Mwiria argues, the emergence and establishment of Harambee schools played a role in the 'legitimation of inequality in Kenya' (Mwiria, 1990, p. 364).

To address financing and maintenance challenges, global water sector policies in the 1980s shifted attention to developing low-cost technologies that can be managed by users with minimal external inputs. This resulted in the design and testing of a variety of handpump technologies that revolutionised rural water services in South Asia and Sub-Saharan Africa. As of 2019, there are close to 700 handpumps in Kitui county (Nyaga, 2019). Most of these are Afridev handpumps which became the technology of choice for much of rural Africa, compared to the No. 6 handpump popularised in Bangladesh. By the end of the twentieth century, the government took a 'very hands-off approach to rural water supply in general' (Harvey et al., 2003, p. 8). Government staff at provincial and district water offices were responsible for coordinating water services in their jurisdiction, providing technical support for borehole siting, issuing drilling permits, and providing ad doc support to communities for water quality monitoring, operation, and maintenance (MWR, 1999).

The enactment of the Water Act 2002 marked the beginning of 'socially responsible commercialisation' of the Kenyan water sector, though the focus was predominantly on urban areas. The Act formalised the government's role in coordination and regulation through establishment of the Water Services Regulatory Board (WASREB) (The Water Act, 2002, Section 46), while the responsibilities for day-to-day service delivery was delegated to water service providers (The Water Act, 2002, Section 55). Water service providers could be commercial companies, NGOs or private entities, licensed to operate in a designated area, and are monitored by WASREB against sector guidelines and standards. Licenses are, however, only applicable for person(s) supplying more than 25 m^3 of water per day for domestic purposes, with schools, healthcare centres or other institutions serving their own occupants being exempted (The Water Act, 2002, Section 56).

In Kitui, there are two such water service providers – Kitui Water and Sanitation Company (KITWASCO) and Kiambere-Mwingi Water and Sanitation Company (KIMWASCO) – which serve a third of the county's population through metered piped water connections on-premises and a network of public kiosks, mostly concentrated in Kitui and Mwingi towns. In fact, almost all the 91 water service providers regulated by WASREB operate in urban growth centres where the population density makes them more commercially viable. Other than these two regulated water service providers, there are 460 rural piped schemes in Kitui, four out of five of which are supplied by groundwater drawn through boreholes using a combination of grid electricity, solar energy, or diesel-operated generators (Nyaga, 2019).

The adoption of a new constitution in 2010 overhauled the country's governance structure, forming 47 new county governments with the aim to decentralise political power, public sector functions, and public finances and ensure a more equitable distribution of resources among regions. By explicitly acknowledging access to safe water in adequate quantities as a basic human right (Article 43), the constitution marked a departure from the market-oriented principles of water as an 'economic good' and set the foundation for extending services to rural areas that are not commercially viable. Under the Water Act (2016), which repeals the earlier act, county governments are now responsible for providing water services within their jurisdiction (The Water Act, 2016, Section 77). County governments are encouraged to contract private entities, community groups or NGOs to manage and operate rural water systems, while trying to embed these small-scale service providers within the sector's regulatory framework through various arrangements with existing water service providers (WASREB, 2019).

Despite multiple reshuffling of ideologies, investment modalities and governance arrangements, the rural water sector till date suffers from two major challenges – financing of new water supply infrastructure and sustainable operation and maintenance mechanisms. Investments in rural water infrastructure have mostly been driven by bilateral and multilateral aid programmes, whether the funds are channelled through government ministries or via project with local NGOs. The

Water Act 2002 mandated the establishment of the Water Services Trust Fund[1] as a pooled funding mechanism drawing on government, donor and private sector funding to target investments in disadvantaged areas. The Water Services Trust Fund's legal mandate limits its funding to formal water service providers which largely ignores the 80 per cent of Kenyans who live in rural areas outside the provision of water service providers. In Kitui, the Water Services Trust Fund has invested in large storage tanks to serve piped schemes in Mwingi and Kitui towns. However, the design specification for the motorised pumps to lift the water to the tanks was inaccurate and not one drop of water has been lifted to these elegant white elephants in the Kitui skyline.

A decade has passed since devolution. With two national elections, several major droughts, and a global pandemic, county governments are slowly developing Water Bills to chart their own pathways to water security (Koehler et al., 2022). While policy documents have progressively encouraged private sector engagement in rural water sector, this is limited in practice both in terms of capital investments and operation and maintenance services. Other than the two regulated water service providers, piped schemes and handpumps are managed by communities through elected representatives or Water Management Committees, individual schools, healthcare centres or churches.

4.4 Seasonal Dynamics of Source Choices and Water Quality

Despite a century of institutional reforms, trying to reallocate responsibilities of infrastructure financing and operational sustainability among the state, user communities, private sector and international donors, it is uncertain to what extent the daily water experiences of Kenya's rural populations have improved. The scene of men, women, and donkeys on the dry riverbed questions whether millions of dollars of development aid succeeded in bringing water to the people. While the availability of potable piped water services 24/7 within the dwelling is a norm in developed countries, it still remains a distant reality for the scattered settlements in Kitui county.

Our 2018 survey of 1,400 households in Mwingi-North reflect the wide range of water sources that people identify as their 'main source of drinking water'. Four out of ten households reported dry riverbed scooping as their main source, followed by handpumps, river, piped schemes/ kiosks, and earthdam being used between one to 2 out of 10 households. But as in Khulna, a focus on main source is eclipsed by the seasonal shifts in water sources – a common phenomenon among

---

[1] The Water Services Trust Fund was renamed the Water Sector Trust Fund after the enactment of the Water Act 2016.

water-stressed rural populations. While the onset of monsoon in Khulna allows people to shift to rainwater conveniently harvested from own roof catchments, the rains in Kitui drive people towards unimproved community sources, mainly earth-dams and surface flows in seasonal rivers, which may not necessarily be closer to home. A week after the last rains, when the surface flows trickle down to the sub-surface, there is a sharp increase in dry riverbed scooping. With each filled jerrycan weighing 20 kg, and each donkey being able to carry up to four jerrycans per trip, the amount of water collected is often limited.

Kasembi Mwinzi is a middle-aged woman living with her two teenage sons and one daughter in Kyuso ward of Mwingi North subcounty. Her husband passed away a few years ago. She sells sand or crushes stones for construction work. She gets water from a kiosk, for which she pays KES 2.5 per jerrycan[2] (USD 1.25 per m^3), and also through scooping from the Kamuwongo River. The river is seasonal but the water table is high. Kasembi goes to the kiosk when she has money, because the kiosk water is better quality than that of scooping. She does not have her own donkey and borrows her neighbour's one. While she does not pay for the donkey she sometimes helps the neighbour in fetching water. When she gets the donkey, she collects 8 jerrycans; however, when she has to carry on her back, she gets only 4 and makes multiple trips. Hence, she reduces water use for laundry and livestock on those days. Only on one occasion she used a private handpump. The owner only allowed her once, because she didn't have a donkey and couldn't go far.

While water in Kamuwongo River can be accessed close to the surface all year round, in most seasonal rivers the water table falls as the dry-season progresses, creating a need to shift to shallow wells dug manually along the riverbeds. Digging and maintaining wells is a labour intensive and time-consuming process. Social relations and affordability are important mediators of accessing wells, and those without own wells often end up paying a high price at the peak of the dry season. Well owners often demand full subscription fee before the start of the season as a precautionary measure against those who tend to shift from one well to another leaving payments due. However, those with good personal relations with the well owner can negotiate to pay in instalments (Bukachi et al., 2021).

Kasyoka Mwangangi lives with her three children and husband in Tseikuru, another ward of Mwingi-North subcounty. She runs a canteen at the town, while the husband does casual work, like construction. She normally fetches water from a water tank near her shop at KES 10 per jerrycan (USD 5 per m^3). The tank is filled up by a tanker truck every few days. When there is no water at the tank, she goes to a private hand-dug well where she buys water at KES 6 per jerrycan (USD 3 per m^3). She prefers the well water compared to the tank, as the latter is often salty and the tank is dirty. But to go to the well she needs to borrow a donkey from her neighbour, and also close her shop for a few hours. Sometimes, when she is really busy and there is no water at the kiosk, she asks a vendor to fetch water

---

[2] Exchange rate USD 1 = KES 100 (as of 2018 when data was collected).

for KES 20 per jerrycan (USD 10 per m^3). She usually needs six cans a day, of which three are used for her canteen, which is adjacent to her home. In September, her water needs were particularly high as she was constructing her house.

In Kitui, it is quite common for men to stay away from home for weeks for paid employment in urban centres within or outside the county. This effectively makes the woman the head of the household, leaving her responsible for the farm and younger children, while the older ones stay at boarding schools. In times of sickness or other crises, women thus need to draw on their social capital and take help from neighbours for water collection (Bukachi et al., 2021). One of our water diary participants, Grace, gave birth during the study period and was unable to fetch water for a month. Since Grace's husband works in Garissa town, she had to buy vended water and ration her use as the water was very costly (KES 25 per 20-litre jerrycan or USD 12.5 per m^3). When the baby was a few weeks old, Grace borrowed her mother-in-law's donkey and fetched water from a handpump a couple of kilometres away. She prefers this handpump to the one closer to her home as the latter is saline.

Like Kasyoka and Grace, many people tend to avoid groundwater-based sources such as hand pumps and kiosks fed by boreholes due to salinity. Groundwater salinity in Mwingi-North subcounty is generally low in the western highlands of Mumoni and Tharaka, where the geology is dominated by quartzites, biotite, and hornblende gneisses. Colluvial deposits and red soils in the low-lying Ngomeni and Tseikuru areas towards the east tend to have higher salinity. Analysis of water quality of hand-dug shallow wells (less than 20 m deep) along the seasonal river-banks, boreholes with handpumps (20–100 m deep) and boreholes with submersible pumps (more than 100 m deep) showed that water salinity in 60–75 per cent of the boreholes exceeded the upper limit for drinking water compared to only 13 per cent for shallow wells Wadira (2020).[3] Those living close to saline boreholes, thus, prefer surface water sources for drinking and domestic use, while using the groundwater sources for livestock.

Our water diaries captured these spatial and seasonal dynamics of water source choices and expenditures (Figure 4.3). On average, households used about four different source types in a given year. The shifts from groundwater to rain-fed surface water sources were more pronounced at the start of the short rains (that is, the first week of December in our study) than the long rains (that is, the second week of April). This highlights the cumulative impact of the prolonged dry-season preceding the short rains, measured in terms of the number of days with zero rainfall.

---

[3] The study was conducted in February 2019 and involved 8 shallow wells, 17 handpumps, and 17 boreholes with submersible pumps in Mwingi-North subcounty. Salinity was measured as electrical conductivity, with 2,000 µS/cm being the recommended upper limit for human consumption.

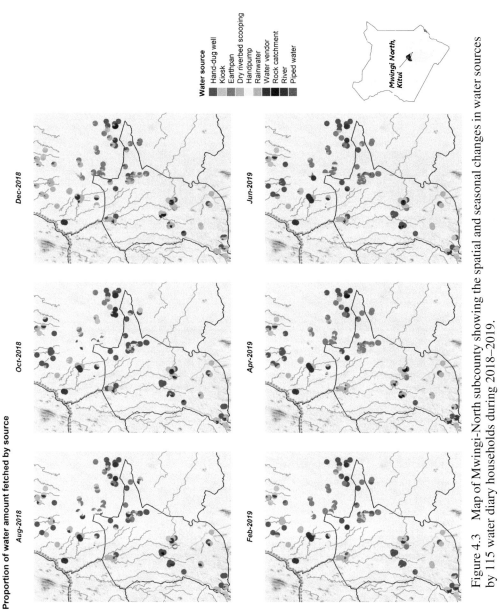

Proportion of water amount fetched by source

Aug-2018   Oct-2018   Dec-2018

Feb-2019   Apr-2019   Jun-2019

Water source
- Hand-dug well
- Kiosk
- Earthpan
- Dry riverbed scooping
- Handpump
- Rainwater
- Water vendor
- Rock catchment
- River
- Piped water

Mwingi North, Kitui

Figure 4.3   Map of Mwingi-North subcounty showing the spatial and seasonal changes in water sources by 115 water diary households during 2018–2019.

There is also a high degree of spatial clustering driven by proximity to different source types and their water quality. For instance, those in Kyuso town use the kiosks throughout the year as these are supplied from the nearby rock catchment or other surface water reservoir, and therefore, have lower salinity than those with motorised boreholes (Figure 4.4).

Such seasonal dynamics were also observed across the 1887 day and boarding schools in Kitui county surveyed in 2019 (Hope et al., 2021b). Three quarters of the schools used two or more water sources, with rainwater being the main source for 30 per cent of schools, followed by piped water on-site (22 per cent), and

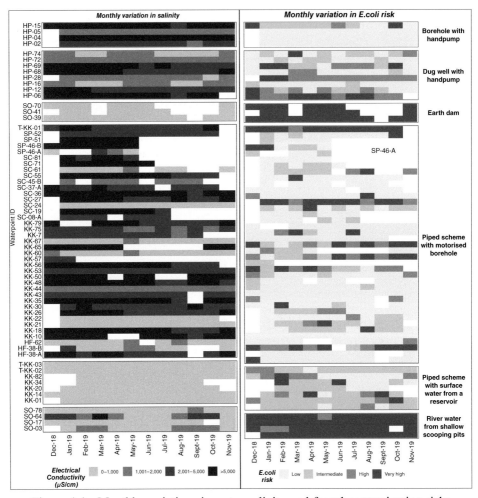

Figure 4.4   Monthly variations in water salinity and faecal contamination risks by type of source in Mwingi-North subcounty. (Designed by author using data from Nowicki et al. 2022. Missing datapoints refer to instances where the sources have dried up, closed operations, or become non-functional.)

vended water (18 per cent). Storage capacity constrained use of rainwater, with 61 per cent of schools having storage capacity of a month or less. Usage of vended water followed the bimodal rainfall pattern, rising from 65 per cent to 95 per cent of schools between July and September, with an estimated annual expenditure of over USD 100,000 across the county.

Depending on the source type, households and schools face high risks of pathogen and chemical contamination (Figure 4.4). A water quality monitoring study in Mwingi-North, led by Nowicki et al. (2022), found very high levels of *E. coli* in all earthdams and dry riverbed scoop holes. High contamination was also detected in a number of piped water schemes and handpumps, questioning the inherent assumption of these technologies being improved and safe for consumption. Very few piped schemes in Kitui treat the water before supply, and in absence of any water quality monitoring, the health risks to users remain uncertain.

4.5 User Payments and Cost Recovery

The choice of water source has a direct impact on household water expenditures. We analysed the weekly variations in water expenditures among our diary participants and identified four distinct expenditure groups (Figure 4.6). Households like Kasembi's, which belong to 'no/low expenditure' category, mostly use free sources like dry riverbed scoopholes or own hand-dug wells. In contrast, households in 'high regular expenditure' category, such as Kasyoka's, fetch water from others' hand-dug wells or water vendors, incurring an annual median cost of USD 167. In between these two extremes, are the 'moderate regular expenditure' ones with a high proportion of water sourced from kiosks and the 'seasonal expenditure' ones who tend to switch from low-cost earthpans in the wet season to high-cost private hand-dug wells in the dry season. While median costs for 'moderate regular expenditure' and 'seasonal expenditure' categories are similar (USD 63 and USD 58 respectively), the distribution of expenses across the year is relatively uniform for the former (Figure 4.5).

There are no significant differences in the amount of water fetched among these four groups. The mean water consumption combining drinking, domestic and productive uses is 4 m^3 per household per month, with one in 10 households consuming 2 m^3 per month. This equates to 22 litres per capita per day, which falls below the WHO's recommended standard of 50 litres per capita per day for basic health and hygiene (Howard et al., 2020). For reference, an individual in a European city is likely to consume 150 litres per capita per day (EurEau, 2020), often rising to 250 litres per capita per day in the United States.

These water source choices and payments raise concerns around 'affordability' of water services. Though affordability is an essential criterion for ensuring

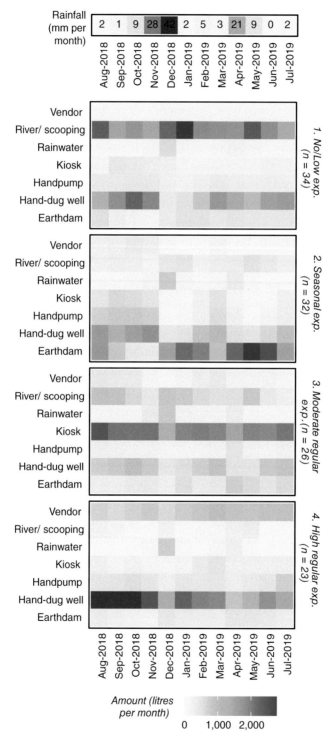

Figure 4.5   Monthly variation in amount of water fetched from different sources and water expenditures for households in four 'expenditure categories'.

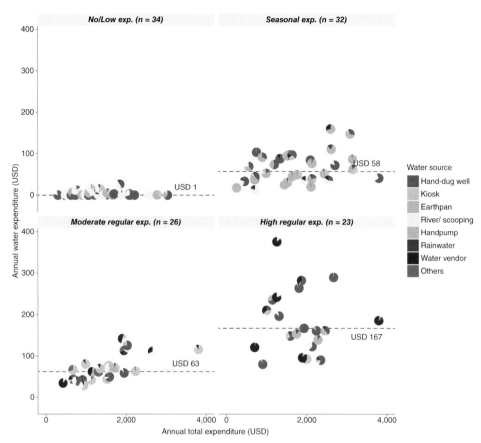

Figure 4.6   Household annual water and total expenditures grouped by 'water expenditure categories'. (Each pie chart represents one household, with the colours reflecting the share of total amount of water fetched by source. Water expenditure categories were derived through cluster analysis of household monthly water expenditures. The dashed lines show the median annual water expenditure for each category.)

the human rights to water and SDG target 6.1, there is little consensus on what it means and how it should be measured (WHO/UNICEF, 2020). While economists have proposed a threshold of 3–5 per cent of household income or expenditure for affordability (UN, 2010), such metrics are inadequate and inaccurate for household water expenses in multi-source settings like Kitui or Khulna. For indicators like quality, quantity, or accessibility, risks are often defined regardless of source or context – for instance, drinking water should be free of faecal contamination, maintenance of basic hygiene requires at least 50 litres of water per person per day a person, or sources need to be located on-premises for safely managed services and within 30 minutes of walking distance for basic services. However, affordability eludes simple definitions and needs to be contextualised in relation to other risk factors like quality, quantity, or accessibility, and costs of other goods and services. Measuring affordability may be straightforward for an average household in the UK spending 1 per cent of their annual household expenditures for 24/7 access to potable water inside dwelling. But for households like Kasembi's, who incur no monetary expenses yet spend several hours a day to fetch a few jerrycans of unsafe surface water, such unidimensional monetary metrics may wrongly signal water services being 'affordable'.

The human rights for water also explicitly states that paying for water must not limit people's ability to acquire other basic goods and services (UN, 2015). In Kitui, food and education are two major expenses, accounting for 32 per cent and 13 per cent of household's overall expenditures which average at only USD 117 per month (or less than USD 1 per person per day). Living on such tight budgets without stable monthly income flows also mean that people need to prioritise their expenses, as well as their time allocated to productive work versus water collection responsibilities. Education expenses constrain budgets in January and September, with some households temporarily reducing the amount of water purchased to balance their budget. Unlike water expenditures that vary widely between households and across seasons, food expenditures exhibit a remarkable steady pattern, peaking only during festival periods. This reveals the difference in market structures between two basic necessities, where demand for food staples is similar across consumers but demand for paid water services vary significantly as part of household needs can be substituted with unpaid sources.

The seasonal dynamics of water source choices have significant implications for financial sustainability of rural water services. Our findings in Kitui mirror previous studies in Africa and Asia where the use of handpumps and piped water schemes in rural areas were found to decrease by 20–30 per cent during wet seasons (Armstrong et al., 2021, Elliott et al., 2019, Thomson et al., 2019). Given that nine in ten piped schemes in Kitui administer a pay-as-you-fetch tariff, charging USD 0.03 per 20-litre jerrycan (USD 1.5 per m^3) on average, the fall in demand translates to decreased revenues, with one in five piped schemes closing operations during the wet season for not being able to cover operation and maintenance costs (Nyaga, 2019). Volumetric data from 2018 to 2021 for 32 piped schemes in Mwingi-North also illustrates these behavioural dynamics, whereby water supplied slumped after the rains started in March and November respectively (Figure 4.7).

In fact, demand shifts are not just driven by seasons but vary on a weekly or even daily basis depending on localised rainfall. Data from rural piped schemes in Ghana, Rwanda, and Uganda analysed against localised rainfall transitions show that if wet seasons are consistent, operators are more likely to experience seasonal revenue reductions regardless of whether the connections are on or off premises

Monthly variations in water supplied across 32 piped schemes

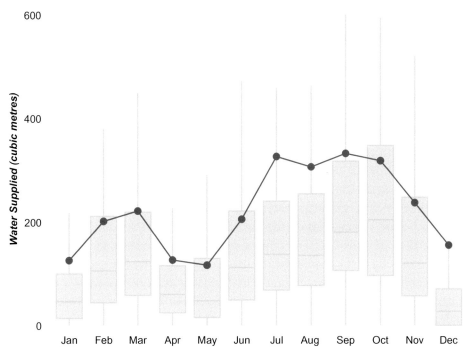

Figure 4.7   Boxplot showing monthly variations in water supplied across 32 piped schemes in Mwingi-North during 2018–2021, with red line showing the mean values. The chart highlights the drop in piped scheme usage during the two rainy seasons (Data source: FundiFix).

(Armstrong et al., 2022). In other words, if the rainy seasons are interrupted by short periods of dry days, people are less able to buffer their daily consumption with rain-fed sources and hence, tend to use and pay for piped services more consistently. These intra-seasonal variations are becoming particularly important in the face of rainfall uncertainties driven by rising global temperatures. Our analysis of school water supplies in Kitui illustrate the widespread vulnerabilities for 400,000 pupils without professional management and monitoring of water supplies.

The way payments are collected also affects how much people pay for water. While piped schemes generally implement a pay-as-you-fetch tariff, payment structures for handpumps vary across individual waterpoints. Some may administer a flat monthly user subscription fee, while others collect contributions as and when needed. Pay-as-you-fetch payments may generate more revenue than flat fees collected periodically, but they can also lead to more seasonal use of multiple water sources. Multiple studies have confirmed that people are less likely to use and pay for water if the source is far from their homes, especially if there are other water sources nearby. Factors like reliable service delivery, and perceived water quality can encourage people to pay for water. Armstrong et al. (2021) argues that temporarily shifting from pay-as-you-fetch to monthly or flat fees during periods when domestic water demand falls or rural incomes are reduced may foster affordable access while maintaining a lifeline of revenue to protect local service providers.

Cost recovery is an essential driver for timely repair and maintenance of waterpoints, as unlike most urban utilities, rural water services are not subsidised. The sector norm is such that the user payments should be sufficient to cover the costs of system repair and maintenance by local private technicians or scheme employees, while the county government can support major assets replacement and network extension. Non-functionality is particularly high for mechanised sources like piped schemes and handpumps, with half of these waterpoints in Kitui found to be non-operational during the 2019 infrastructure audit (Nyaga, 2019). Functionality rates also vary by type of management model, with evidence from Kwale county along the coast showing that handpump downtimes are much higher for community-managed waterpoints (36 days) than for those managed by schools or healthcare centres (20 days) (Koehler et al., 2018). The community-based management model is based on an 'egalitarian' risk-sharing culture, where financial risks of waterpoint repair and maintenance are meant to be equally shared by users. However, only half of community-managed waterpoints have regular user payments, while the rest rely on fund collection upon waterpoint breakdown.

The limits of community-based management have led to experimentation with alternative operation and maintenance models in parts of West and Sub-Saharan Africa, though most are limited in scale (McNicholl et al., 2020). FundiFix, a professional maintenance service delivery model, has been active in Kitui and Kwale counties since 2015. The FundiFix model reallocates responsibilities for operational risks from voluntary community organisations to a social enterprise guaranteeing repairs within a few days (REACH, 2016). When managed by user communities, broken handpumps and piped schemes take weeks to months to be repaired, with households facing an additional cost burden of USD 0.43 per day when fetching from alternative sources (Foster et al., 2022). This downtime is reduced to two days when maintenance services are professionalised, generating significant social and economic returns. These benefits can be further optimised if users continue to use piped schemes and handpumps during wet periods, instead of shifting to unimproved sources.

Exploiting the observed operational and financial data from FundiFix, our colleagues modelled the potential impacts for Kitui county if professional service providers managed all water supply infrastructure (Chintalapati et al., 2022). The results estimate functionality would increase from 53 to 83 per cent with a 67 per cent increase in water production due to higher reliability. The financial implications for government and donors are also appealing with a 60 per cent reduction in the costs of major repairs due to preventive and rapid maintenance services. While water user payments currently cover 15–20 per cent of FundiFix's local operation costs, the dramatic improvement in results can crowd in new funding sources.

Since 2016, the Water Services Maintenance Trust Fund has provided a results-based contract to FundiFix as policy experiment in Kitui county.[4] Initially, research funds from UK Foreign and Commonwealth Development Office were used to incubate the model – demonstrate how results-based contracts may work and how much they would cost. The positive results have attracted corporate funding from national and international partners leading to the majority of subsidy being paid by these partners with minority support from traditional donors (WSMTF, 2023). This presents evidence for Kitui county and the National Water Services Trust Fund that this model could provide more water reliable and safely at lower cost than current practices. Time will tell if there is political leadership and commitment to make these positive changes at scale in Kenya. As we have seen in Bangladesh (Chapter 3), government partners have collaborated in testing and now scaling up a major results-based funding programme for schools in the coastal zone with a government commitment of 50 per cent of the results-based contract from 2024 to 2030. At a cost of less than USD 1 per person per year, it seems a good investment to build the education and health of the next generation.

4.6 Conclusion

It would be simple to conclude that water insecurity in Kitui County is a function of a more unpredictable and punishing climate as rainfall patterns change and temperatures increase. The diaries offer a more nuanced interpretation as the daily water use practices reveal insights into culture and behaviour that is moderated by historical and evolving issues of water governance in managing, coordinating, and delivering safe drinking water services. We consider three findings which may have wider implications for similar remote dryland areas in Africa.

First, the diaries reveal that households choose different water supplies across the seasons. Women will walk past a new kiosk with safe water in favour of a traditional well with uncertain water quality. Even households with 'high' water expenditure will blend water bought from vendors delivered to the house with unimproved well water. Of note, is that despite the variation in sources chosen and money spent, the average water use per month is around 4,000 litres per month,

---

[4] See www.kituiwaterfund.org.

or just over 20 litres per person per day. We find no wealth effect in consumption levels, only in source preference and payment level.

These findings do not align well with global monitoring efforts which assume 'one main water source' or legal guidance on a minimum water quantity. Policy and investments based on the latter seem unlikely to achieve desired outcomes. Further, the low (less than 2 per cent) or no expenditure on water suggests affordability issues may be compounded by cultural practices influenced over time. The likelihood users will value and pay for higher water quality seems limited based on diary behaviour. This poses multiple issues for governance and policy.

Second, governance issues are multi-scalar from the household to the community, and from the district to the county. Guaranteeing reliable drinking water through a professional service provider has reduced repair times from over 30 days under community management to less than 2 days under a professional service provider (FundiFix). Community uptake has been voluntary, slow, and uneven. Cost recovery is challenging, and a subsidy is required to operate the service effectively. Analysis suggests county uptake of a professional service delivery model would reduce county government expenditure, guarantee reliable water, and increase production (few service breaks) which may also increase revenue. This requires county leadership and donor cooperation.

For decades, rural water schemes have been funded by government or donors and then handed over to communities to fail or be abandoned within a few years. Inevitably, communities have had to find alternatives making their own investments in wells, rainwater harvesting or buying vended water. This has increased the availability of sources though often of uncertain water quality or proximity. Changing this behaviour is unlikely to be quick or straightforward. Droughts create significant hardship leading to expensive tanker trucks being required, draining resources in a short-term fix to a systemic problem. Kitui County Government has introduced regular sector meetings and slowly advanced a common policy and strategic framework. However, large projects from external funders regularly divert from a common plan leading to further wasted resources and embedding a cultural of self-dependency and non-payment.

Third, public facilities such as health clinics and schools face similar problems without a regular revenue stream from paying customers (Nyaga et al., 2024). Services are extremely poor leading to high costs for around one-third of county schools paying over USD 100,000 in dry periods after rainwater tanks are used. Unlike communities, schools and clinics are notionally under national management due to policy ambiguity. Despite the increased risks from limited functioning waterpoints during COVID-19, no funded plan has emerged to ensure safe drinking water services are available for these vulnerable groups. While efforts focus on community water supplies, the children in schools and patients in clinics are excluded. Girls are particularly affected both in lower attendance rates without water for menstrual hygiene management and cultural pressures to collect household water when waterpoints fail. Cycles of marginalisation are reinforced causing avoidable harm to individuals, their families, and society at large.

Kitui's water practices today are a reflection of cumulative decisions from the colonial period until today. Structural inequalities and weak accountability have progressively increased water insecurity. There is no doubt that the changing climate will aggravate the hardship for most rural people. Policy and investments have not had sustained impacts for decades. Donor projects fail without any accountability, national government has transferred a legacy of failure to newly elected county governments. Local people have learnt how to survive with limited trust in external interventions. The seemingly irrational water use behaviour reflects generational knowledge and disappointment. Professional service delivery has shown what is possible through achieving scale depends on leadership by the elected Governor and multi-donor cooperation. With limited capacity and resources, the challenges are significant. However, Kitui has incubated an effective model to guarantee drinking water, with progress to include water safety. The opportunity rests with government and donors to support and cooperate in avoiding the mistakes of the past to deliver a water secure future for all.

5

Small Towns in Arid Lands

Unreliable Piped Water Services and Flash Floods in Lodwar

5.1 Introduction

About 600 km northwest of the Kenyan capital, Nairobi, lies the country's largest, driest and poorest county – Turkana. With daytime temperatures hovering around 35°C and virtually no rainfall for most of the year, the stark aridness of the landscape is characterised by a few hardy trees and red dusty expanses where seasonal gullies and channels are the only fingerprints of rare but intense rainfall events. Lodwar, the largest town and headquarters of Turkana County, is nestled between two rivers and overlooked by black volcanic hills. One of the two rivers is the Turkwel, the only perennial river in the county that originates from southern Uganda, flowing north from Mount Elgon and draining into Lake Turkana – the world's largest desert lake. Boreholes dug along the Turkwel serve the town's 83,000 people through a piped water network, though supply is highly unreliable and unequal across the town's neighbourhoods (Tanui et al., 2020). The other river is the Kawalase, a seasonal river or 'laaga' as they are called, which has earned the nickname of the 'river of death' for its dangerous flash floods that engulf cattle, vehicles, and anyone unlucky enough to be caught in their violent and sudden path.

While Turkana also has a bimodal rainfall pattern like Kitui, rainfall in the Turkwel basin exhibits high spatial variation, ranging from 900 to 1,700 mm per year in the upstream areas to only 200 mm in the downstream section where Lodwar is located (Hirpa et al., 2018) (Figure 5.1). Droughts are common in these arid and semi-arid landscapes in the Horn of Africa and occur due to a complex interaction of hydroclimatic factors resulting in consecutive failed rains during the October–December 'short rains' season. Two of the most severe droughts of the twenty-first century occurred in 2017 – the year we started our fieldwork in Lodwar, and in 2022 – the year we are writing this chapter. And in between these years, Kenya recorded some of the wettest years in recent history, raising alarms

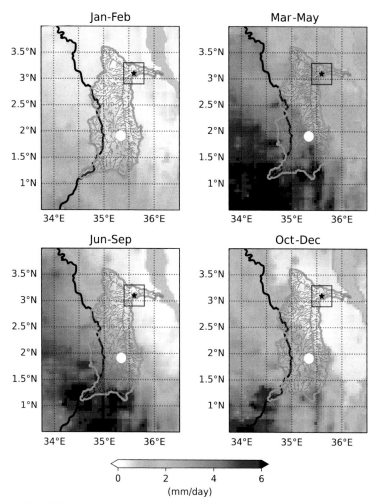

Figure 5.1 Charts showing spatial variations in rainfall in the Turkwel River basin (orange boundary) across different months. (Star shows location of Lodwar town, and white dot shows Turkwel Gorge dam). Map drawn by Ellen Dyer using rainfall data from 2016 to 2022 available from the Climate Hazards Group InfraRed Precipitation with Station data (CHIRPS).

of widespread flooding in late 2020 as the Turkwel dam was at risk of overflowing for the first time since its construction in the late 1980s (Macharia, 2020).

Turkana has always been a land of extremes, not only in terms of climate but also in its geopolitical importance. Inhabited by nomadic pastoralists who periodically fight over scarce grazing grounds, Turkana has long been treated as an area of low economic value and unworthy of development investments, both by the colonial and post-independence regimes. The result is that today the county has the highest poverty rate in Kenya, with four out of five people living in absolute poverty, unable to meet the very basic needs for food and drinking water. Adult literacy is 20 per cent and less than half of school-aged children are enrolled in primary school. Following a century of social exclusion, it is not surprising that the locals have a sense of separation, and they often joke that they are travelling to Nairobi to visit Kenya.

However, in the past decade, Turkana, and Lodwar in particular, came to the limelight with the discovery of oil in 2012 and two major aquifers in 2013. At the same time, the ongoing construction of the LAPSSET (Lamu Port – South Sudan – Ethiopia Transport) corridor – an ambitious infrastructure development project for connecting 160 million people across four East African countries – has elevated the strategic importance of Lodwar as it links Kenya with neighbouring South Sudan and Uganda (Schilling et al., 2016). To what extent the benefits of these developments will be passed on to the Turkana people is unknown, but for now the hopes of better economic opportunities and living standards have made Lodwar a desired destination for many rural migrants. International aid organisations, construction companies, and businesses are increasingly making a base in Lodwar. Thus, what was once a 'closed district' under British colonial administration is now well connected to the world, with multiple direct flights between Lodwar and Nairobi every day.

But life in Lodwar is not easy and one of the many challenges is water. Despite being cradled by two rivers and close to a reportedly large groundwater reserve, Lodwar remains a thirsty town due to significant institutional and operational risks associated with the town's piped water supply. The town's growing population, having almost doubled between the 2009 and 2019 census, and the planned irrigation schemes upstream along the Turkwel River are driving up demand for groundwater resources. At the same time, frequent occurrence of severe droughts is constraining water availability in the basin by reducing run off and groundwater recharge. The problem of too little water also coexists with that of too much water, as a section of the town's residents live in fear of being swept away by flash floods often occurring in May or October. The decision to relocate to safer grounds has opportunity costs. Life outside the town boundary is challenging with limited income sources and no water, education, or healthcare facilities. For those who decide to move away, the 'river of death' ultimately becomes their lifeline. The scoop holes and shallow wells on the riverbed may not provide safe water, but it becomes the only reliable and affordable source.

Lodwar offers a unique opportunity to understand how people negotiate between two different water risks – exposure to flash floods and drinking water services – while trying to make a living within an urban space. While risks emerge from the interplay of environmental, institutional, and economic factors, who is at risk and how they navigate them depends on their intersectional identities – their gender and age, where they are from, their family and relationships, their education, and socioeconomic status. In this chapter, we delve into the life histories and water diaries of the people of Lodwar to explore how experiences of risks are differentiated across social and spatial axes. We embed these stories within the broader landscape of development activities and climate change in the region to discuss how risks are likely to play out in future.

5.2 Drought, Destitution, and Development

In 2022–2023, as the world continued to recover from the COVID-19 pandemic and the economic repercussions of Russia's invasion of Ukraine, countries in the Horn of Africa endured the worst drought in four decades. Major rainfall deficits for five consecutive rainy seasons since late 2020 led to drying up of soil moisture and waterways, causing crop failure, widespread death of livestock and severe hunger, with 25 million people in Kenya, Ethiopia, and Somalia facing extreme water insecurity (OCHA, 2023). The drought was the result of a prolonged multi-year La Niña event and a concurrent negative phase of the Indian Ocean Dipole, both of which caused dry weather and high temperatures in East Africa by pulling away moisture towards Southeast Asia. While the El Niño-La Niña cycles and the Indian Ocean Dipole – periodic fluctuations in ocean temperatures – are natural phenomena driving global climate and precipitation patterns, overall rise in sea surface temperatures caused by climate change is thought to amplify the intensity of these events in recent years.

Droughts are a defining characteristic of these arid and semi-arid regions, and have been instrumental in shaping the livelihood strategies, population dynamics and development trajectory of Turkana, with implications for present-day water insecurities in Lodwar. For centuries, the nomadic pastoralists of Turkana have used their knowledge of the environment and social networks to strategically navigate these fragile landscapes in search of water and pasture for their livestock. To cope with prolonged dry spells, pastoralists adopted flexible herd management practices, including herd diversification, herd-splitting and clear sex- and age-set-based division of labour (McCabe, 1990). Persistent droughts exacerbated livestock raids and ethnic conflicts over access to dry-season pasturelands, notably with the Merille tribe in bordering southern Ethiopia and the Pokot tribe in West Pokot and Baringo Counties in Kenya and the Pokot District in bordering eastern Uganda.

Throughout the twentieth century, however, the resilience of Turkana herders has been eroded through a combination of geopolitical and environmental pressures, many of which can be traced back to the exploitative policies of the British colonial administration (Abdullahi, 1997). The Turkana people, who actively resisted British domination of their homelands, were considered a nuisance to the European farmers in neighbouring Trans Nzoia County (Nicholas, 2018). At the same time, the pastoralists' primitive mode of production was of little economic value to the colonial regime. Thus, to 'contain' the Turkana pastoralists, the British launched a major punitive expedition, confiscating and slaughtering over a quarter million livestock, and deliberately segregating the Turkana people by declaring it a closed district in 1920 (Nicholas, 2018). The free movement of people, which was further restricted by demarcating national boundaries and reserving fertile land for commercial farming by white settlers, prevented the pastoralists' ability to balance seasonal variations in water and fodder availability across grazing areas.

The systemic decline in coping capacities increased pastoralists' vulnerability to droughts overtime, causing many herd owners to move into famine relief camps, food distribution sites, irrigation schemes, as well as small towns and trading centres, one of which is Lodwar (McCabe, 1990). A series of devastating droughts and famines in 1960–1961, 1974–1975, and 1980–1981 spurred the influx of relief and development efforts by the international donor community, with half of Turkana's residents being on relief rolls in the early 1980s (Akall, 2021). To enhance drought resilience and reduce relief dependency, the post-Independence Kenyan government and the donor community promoted shifts from pastoralism to irrigated agriculture and fishing in Lake Turkana (Derbyshire, 2020). Settled farming would also allow provision of basic services, such as clean water, healthcare, and education, that the Turkana people has historically been deprived of. Despite large capital investments, these projects mostly failed to achieve their objectives in the long term, as settled agriculture was a markedly different way of life that the pastoralists found difficult to adopt (Hogg, 1982).

The influx of donor investments in the mid 1980s, however, facilitated the transformation of Lodwar from a remote, dusty, and relatively inconsequential town into a growing economic centre with an array of shops and services. While the poorer cattleless Turkana migrated to Lodwar to seek humanitarian aid facilitated opportunities, herders and fishermen from surrounding villages gathered for trade and created demands for consumer goods. However, as aid money dried up in the 1990s, the drought-stricken pastoralists found themselves in a one-way poverty trap. Having spent more than a decade on the donor sponsored programmes, such as the food-for-work schemes under the Turkana Rehabilitation Project, the pastoralists lost the social networks to (re)establish themselves in the pastoral economy, without being able to establish a firm foothold in the new economic environment either (Broch-Due and Sanders, 1999). Today, Lodwar is home to many of these destitute people who struggle to survive in a cash economy, through meagre earnings from charcoal and firewood sale, basket weaving, beer brewing and casual labour, in the driest part of the entire county.

5.3 An Obituary of a Water Utility

Across Turkana county, surface water sources, including seasonal rivers and shallow wells on dry riverbeds, and boreholes are predominantly used for drinking and domestic purposes. There are about 1,500 boreholes in the county, installed by various organisations over the past four decades – the most prominent of these being the Catholic Diocese of Lodwar. The diocese, which started its journey in the 1960s with the arrival of Irish missionaries, plays a pivotal role in provision of water, education, and healthcare services across the county. Only a handful of towns in Turkana have piped water network, but none as extensive as Lodwar. The responsibility of Lodwar's piped water supply lies with the Lodwar Water and Sanitation Company (LOWASCO) – a private utility operating since 2007 through a Service Provider Agreement with the Rift Valley Water Services Board. However, following the Turkana Water Act 2019, water service delivery in high-density areas will be contracted to one or more urban and rural water companies that will be regulated by the newly established Turkana County Water Department (Turkana County Water Act, 2019) (Figure 5.2).

The piped network is entirely dependent on groundwater from the shallow alluvial and intermediate aquifers recharged by the Turkwel River. Most of LOWASCO's production boreholes operate on hybrid pumping systems of solar and grid electricity, with average daily total abstractions ranging from 4,000 m^3 in the dry season to 5,200 m^3 in the wet season. While the boreholes yield enough

Figure 5.2   Typical dome-shaped huts in Lodwar with a metered LOWASCO water tap protruding from the ground (Photo credit: Sonia Hoque, February 2019).

water to serve the municipality, as of 2019, the network covered only 58 per cent of the population in the service area with 8,000 connections (Olago and Tanui, 2023). Apart from individual household connections, water is supplied through communal points within a compound, kiosks, and a network of both utility and privately owned tanker trucks. The spatial heterogeneity in services is evident from the household survey we conducted in 2017, where 31 per cent of households within LOWASCO's service area reported using kiosks as their main water source, followed by 24 per cent who borrowed or bought from neighbours, 20 per cent who resorted to scoop holes or wells by the river, and only 19 per cent who used piped water within their dwelling or compound (Figure 5.3).

Jessica Ekodos, a 31-year-old mother of three, lives in Namakat village along the Turkwel. 'Life was good' when she moved here with her husband 16 years ago. Her husband worked as a security guard, while she walked door to door, selling Mandazi (Swahili fried bun) and vegetables. They were happy to find a place that is close to the town as well as the river which they used for drinking, washing, and bathing. Although there was no electricity or piped water, people had food to eat. She even took her children to ride the merry-go-round, as she fondly remembers.

A few years later, the village elders wrote an application to urge LOWASCO to connect this area to the piped network. Jessica's house was connected in 2014. But in 2016, they were disconnected due to outstanding bills. While LOWASCO said they had KES 7000 (USD 700)[1] due, Jessica believes it was not more than KES 500 (USD 5). Now they fetch water from their neighbour and contribute KES 100 (USD 1) to their bills.

For Lodwar's residents like Jessica, there are three main complaints about LOWASCO's water service: first, high connection charges which vary based on the household's location in relation to the main distribution pipe; second, faulty meters and discrepancies in billing often resulting in unusually high bills and disconnection due to outstanding payments; and third, an unreliable service meaning the taps can be dry for days or weeks at a time. Whether one is connected or not, water expenditures are usually quite high and unlike Kitui, there are no seasonal pulses in water sources and costs (Figure 5.4). Those fetching most of their water from own piped connections reported an average monthly bill of KES 500 (USD 5), while those getting from neighbours, handpumps or kiosks paid an average of KES 1,600 per month (USD 16) (Figure 5.5). Yet, regardless of source, the total volume of water used was only about 4 m^3 per month per household, equating to approximately 20 litre per capita per day.

Despite households bearing such a significant cost burden for water, more than 40 per cent of LOWASCO's water supplied is unaccounted for, posing a major threat to the sustainable service delivery. A network mapping study in 2018 confirmed that a

---

[1] Currency conversions are based on exchange rates at the time of data collection, averaging at USD 1 = KES 100 during 2018–2019.

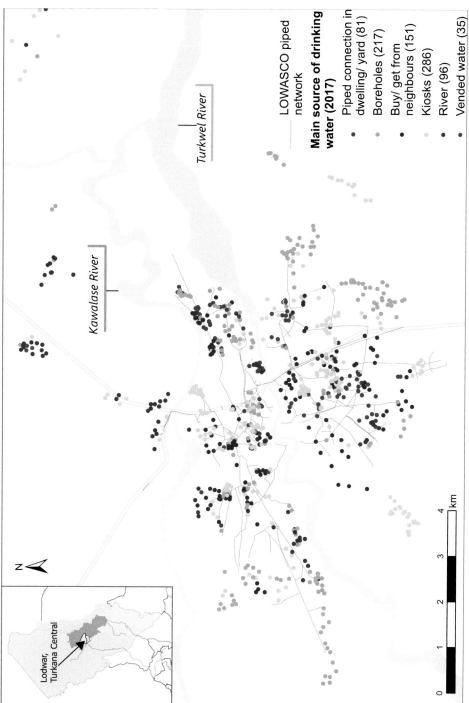

N

Lodwar,
Turkana Central

Kawalase River

Turkwel River

LOWASCO piped
network

**Main source of drinking
water (2017)**

- Piped connection in
  dwelling/ yard (81)
- Boreholes (217)
- Buy/ get from
  neighbours (151)
- Kiosks (286)
- River (96)
- Vended water (35)

0   1   2   3   4
km

Figure 5.3   Main sources of drinking water reported by households in Lodwar town in 2017.

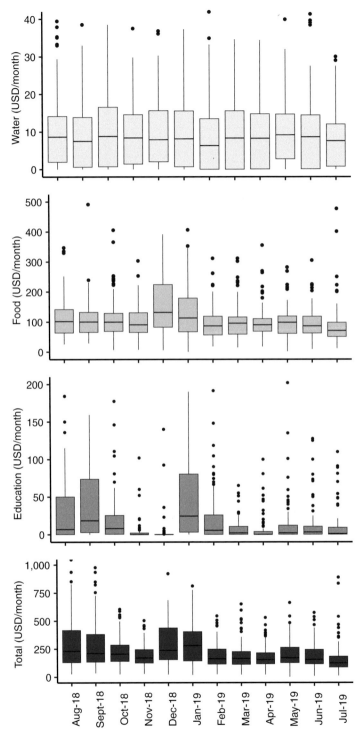

Figure 5.4   Monthly variations in water, food, education, and total expenditures reported by 98 water diary households during 2018–2019. Water expenditures remain relatively stable throughout the year, with food expenditures peaking during Christmas (December 2018) and educational expenditures peaking in beginning of term (September 2018 and January 2019).

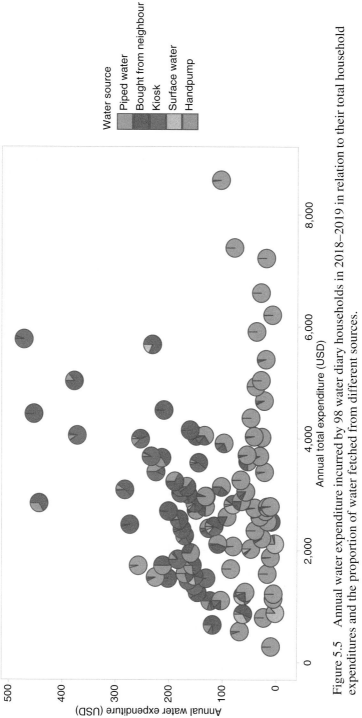

Figure 5.5   Annual water expenditure incurred by 98 water diary households in 2018–2019 in relation to their total household expenditures and the proportion of water fetched from different sources.

considerable amount of water is lost due to overflows of storage tanks with limited capacities, and leakage along the main pipelines (Maxwell et al., 2020). In 2019, the Office of the Auditor-General released a scathing report of the utility criticising the lack of a proper Board of Directors or Corporate Secretary, and revealing stark anomalies across the scale of LOWASCO's operations (OAG, 2019). Improper representation, financial inaccuracies, discrepancies in provided cash flow statements, lack of transparency over assets from the Rift Valley Services Board, and opaqueness over company expenses were documented. The lack of evidence was so great that the Auditor-General struggled in concluding the lawfulness or lack thereof of utilisation of public resources by LOWASCO. And then, in 2021, LOWASCO made headlines when the Kenya Power Company disconnected energy supply for LOWASCO's operations, as it owed electricity bills worth KES 11 million (Etyang, 2021). The residents went weeks without any supply of drinking water, until the county government mediated the situation between the two companies

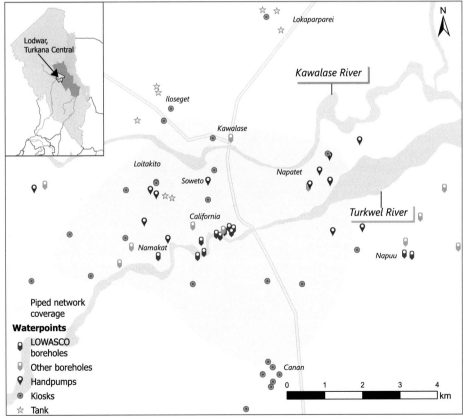

Figure 5.6   Map of Lodwar town showing location of waterpoints (functional at the time of data collection). Data combined from multiple rounds of water supply infrastructure mapping in June 2021, February 2022, and May 2023.

Over the years, a number of NGOs and international organisations have collaborated with LOWASCO to support their operations through diverse initiatives including the installation of solar-powered boreholes and automated water kiosks (Figure 5.6). Since the ATM-operated kiosks are designed to be self-serviced, they do not have any paid operator and engage community volunteers to report faults and prevent vandalism. Unlike other LOWASCO built kiosks, where private operators get to keep part of the revenue, the caretakers of the ATM-operated kiosks were frustrated as they are burdened with the unsatisfactory job of mediating user complaints and delayed actions by the utility. Interestingly, or rather unfortunately, the automated technologies have led to a small-scale black market of water resale, as those with cards can purchase enough water to deplete the storage, creating an artificial scarcity. In fact, two of the ATM kiosk managers reflected that the idea of reselling the water at a higher price was a means of compensating for their time commitments that are otherwise unappreciated.

5.4 Floating Elephants and Sinking Livelihoods

But water supply is just part of the many challenges of life in Lodwar. For people like Jessica, who live in low-lying areas such as Namakat along the Turkwel River and Napatet, Soweto and Loitakito along the Kawalase River, there is constant fear of being washed away by flash floods. These areas are often occupied by the poorest who risk living in flood prone places to benefit from proximity to central business areas where government and NGO offices, and commercial establishments are located.

For Jessica, life became harder when her husband lost his job in 2017. Soon after that there were two successive floods, and the one in October 2019 was the 'mother of all'. It was even more destructive than the flood of Etom (elephant) in 2006 – one that carried away an elephant calf. Since Jessica had three separate houses within the compound, she still managed to save her stove, mattress, and some documents by moving them to the raised house, while the other two were washed away. One good thing was that the waters came in the afternoon. 'My children are safe, but I lost my five goats', she said, while pointing at the Mathenge trees where her goats' shelter used to be.

Jessica, like many others in her village, wants to relocate to a safer place. She has a plot of land north of Kawalase, but any 'rational person with children' will hesitate to move there as the place has no schools or water. The route to the nearest school in Nakwamekwi is unsafe. If they had a motorbike, they could have used it for transportation as well as carrying water.

As Jessica was sat breastfeeding her youngest son, she mentioned that she is now 'living her worst life'. Her husband now does construction work which becomes available once in a while. 'Life is good when you have food. Here you can stay a whole day without eating anything', she added. Given her childcare responsibilities, she is unable to go and look for work. If only she could get some work, they would be able to save money to buy land and relocate to a safer area.

Jessica's story epitomises the struggles and concerns of the poorest people of Lodwar – getting food to eat, education for children, water for the family, and a safe place to live. Lodwar has a relatively high simple daily intensity index, which means the infrequent days that it does rain, it rains a lot. Infrequent rain and high temperatures followed by intense rainfall can trigger flash flooding, a characteristic of the Kawalase, and river flooding which can occur when the Turkwel River exceeds its banks and recedes. The 2019 flood, which Jessica recalled, intensified the town's water crisis, as all the LOWASCO boreholes along the Turkwel River were flooded and remained non-operational for more than a month, as the authorities struggled to source damaged parts that needed replacement (Etyang, 2019).

Relocation to peripheral villages such as Lokaparparei and Iloseget, on the other side of Kawalase River, is a common coping mechanism, though there are multiple factors to consider. Living close to the town centre provides income opportunities through small businesses, like selling vegetables, or casual jobs, like doing laundry or construction. Ajikon Akwar, a 38-year-old mother who moved from Napatet to Lokaparparei, mentioned, 'It is so easy to go without food here. The main income generating activity is fetching woods and palm leaves for weaving baskets. When I am not feeling well or I have chest pain, I stay at home and start weaving baskets. When I am done, I can sell them for KES 20 (USD 0.20) each. I use it to buy maize flour and make porridge for my children, without sugar or salt.' By moving to new areas, people also lose the customer base that they developed over the years. Moreover, when most people in the community are poor, there is little demand for goods and services. Gender often intersects with spatiality to restrict access to employment or social support. Regina Ewoi, a 22-year-old resident of Soweto, whose father left her mother after remarrying another woman, believes that having 'no boys in the family' is a major cause for their distress. 'Boys are outgoing. They mix with different people and get information about job opportunities, bursaries, and aid.'

To be able to relocate, one also needs access to land and money to build a house. In Turkana, most of the land is community owned and bestowed upon the County Council under the Trust Land Act. Individuals do not hold titles to the land they occupy, except in urban areas like Lodwar, where they are given allotment letters as proof that they are occupying the land legitimately. In Lodwar, owing to the high in-migration of outsiders, about 65 per cent of the land occupiers are non-ancestral settlers (World Bank, 2015). While early migrants could claim any unoccupied land, land has become increasingly scarce and those with allotment papers are selling land to property developers. Several factors have contributed to this increased land demand. Since the devolution in 2013, there has been increased demand for housing to accommodate county government staff and officials of non-governmental organisations. At the same time, the upgrading of the 960 km Eldoret–Juba road linking Kenya and South Sudan attracted investments in satellite colleges, petrol stations, bars, retail shops and modern residential housing in small towns like Lodwar (Business Daily, 2016). Thus, while one may temporarily settle on uninhabited land borrowed from a friend or neighbour; there is risk of being evicted if the owner finds a suitable buyer.

Even when one gets access to land, lack of finance to build a house can be a deterrent to relocation. Development organisations like Red Cross have occasionally supported people in getting land and building houses, only to find those houses being abandoned after some time. As Jessica narrated, lack of income, absence of schools and healthcare facilities, and lack of drinking water were reported as the major constraints of life in these villages. Since these are outside the piped water network, residents depend on either water tanks or the Kawalase River. Of the five water tanks we mapped in Iloseget and Lokaparparei, one was provided by the county government and four were privately owned by residents who saw a business opportunity in selling water (Figure 5.7). In all cases, the tanks were filled by

Figure 5.7   Private water vendor selling water for KES 30 per 20-litre jerrycan (USD 1.5 per m^3) in Lokaparparei, 4 km north of Lodwar town (Photo: Waterpoint Survey, July 2021).

tanker trucks carrying water from LOWASCO boreholes. While people are content with the quality of the tank water, supply can be very unreliable.

Esther Lotieng owns and operates a water tank in Lokaparparei, a village 4 km north of Kawalase River. When she came to this area, she realised that people do not have water to drink. She asked her husband to save some money and buy her a tank so that she can earn some money and also have water for her family. She pays KES 5000 for the water bowser (tanker) to bring 5000 litres of water from Moi gardens borehole and sells this water for KES 30 per 20-litre jerrycan (USD 15 per m^3). People in Lokaparparei are poor and cannot afford to buy water. But when it rains, Esther's sale increases as people are scared to go to the river during high flow. Before the Kawalase bridge was constructed, the water supply was unreliable. One would receive a call from the truck driver saying that they would not come due to risk of floods.

Bulk sale of water to tanker trucks, whether owned by private vendors or by LOWASCO, is a major source of revenue for LOWASCO. Tanker trucks fetch water from LOWASCO boreholes and sell to construction sites, schools, or community storage tanks outside the service area. Individual households in these areas can purchase water from the tank owners or hire *boda bodas* to fetch water from kiosks or boreholes in the town. While the kiosks sell water for KES 5 per 20-litre jerrycan (USD 2.5 per m^3), households end up paying KES 120 for four jerrycans (USD 15 per m^3), with the KES 100 being charged for transport.

But for the vast majority, vended water is unaffordable, and river is the only source within their means. The dry riverbed of Kawalase becomes a busy area

Figure 5.8   Children scooping water from the dry riverbed of Kawalase River in Lodwar (Photo credit: Sonia Hoque, February 2019).

early in the morning and in late afternoon when the sun is a bit more forgiving. Women and children, and few men, can be seen digging scoop holes to fill up their jerrycans pint by pint (Figure 5.8). Since children cannot be left at home on their own, mothers have to carry young children on their journey to fetch water while older children end up walking long distances alongside. Open defecation along the river amplifies water collection challenges, as new wells need to be dug every day to avoid contamination by human faeces. Water is usually never treated before consumption as such additional tasks seem impossible due to physical exhaustion, coupled with lack of time and resources. As one respondent mentioned, 'By the time we reach home, our children are thirsty. They will be crying "Mama, Mama," so you feel sorry and decide to give the water the way it is. Boiling and cooling the water takes like forever.'

5.5 Governing Groundwater for Growth

While the daily water challenges of Lodwar's residents have changed very little over decades, there have been several significant institutional and infrastructure developments in the past decade which have rejuvenated hopes of a better future. In 2010, Kenya enacted a new constitution, establishing a devolved government with large-scale political, fiscal, and administrative decentralisation to 47 newly formed counties. The devolution, which came into effect after the 2013 general elections, enabled Turkana to take greater accountability and responsibility for its development. From a neglected district largely under the pastoral care of the Catholic diocese, Turkana elected its own governor and county assembly, with access to an equitable share of the national revenue. In 2012, the county's fortunes also appeared to have prospered when the UK listed company Tullow Oil announced the discovery of 750 million barrels of commercially viable oil in Lokichar, about 90 km from Lodwar (Mkutu Agade, 2014). The following year, Turkana seemed to have won the water lottery. Using satellite exploration technologies, deployed by Radar Technologies International, identified the presence of two large aquifers under the county's dry thorny landscape. The Lotikipi Basin Aquifer and the Lodwar Basin Aquifer were estimated to hold about 250 billion m^3 of water, enough to supply the whole country for the next 70 years (Gramling, 2013).

   The lure of two precious natural resources, coupled with the newfound institutional and fiscal autonomy, reawakened the sleepy transit town of Lodwar. With the influx of shops, guesthouses, leisure centres and hotels, the demand for water services have continued to rise and new connections have been added to the network without proportionate increases in supply. Despite the hype surrounding the discovery of aquifers, there has been no progress in actually bringing the water from over 1 km below ground to taps in people's homes. A small-scale drilling project by the government in 2015 identified that the water in Lotikipi aquifer is highly saline, and unfit for human consumption, irrigation or livestock. This sub-dued the hopes of a miraculous solution, and led to conversations with investors about desalination, though the project did not take off as energy costs were deemed higher in lifting and treating the water than any feasible economic usage. Ideas to irrigate food crops or grow cotton are politically popular but financially redundant.

While groundwater is widely recognised as a strategic resource for economic development in these rain-deprived drylands of Sub-Saharan Africa, there is insufficient understanding of aquifer characteristics, including their recharge mechanisms and potential impacts of geogenic and anthropogenic activities. In Lodwar, potable groundwater can be obtained only from recent alluvial and Holocene age sediments of the Lodwar Alluvial Aquifer system (less than 100 m below ground level), as older sediments mostly yield saline water (Tanui et al., 2020). Further analysis of isotopic composition confirms that the aquifer is mainly recharged by surface flows from the perennial Turkwel River and by infiltration of local precipitation during the wet season. This makes the aquifer susceptible to upstream flow regulation by the Turkwel Gorge dam and prolonged droughts, as well as faecal contamination resulting from a lack of sewerage network and municipal waste disposal system (Tanui et al., 2023).

5.6 Conclusion

It is hard to reconcile the image of a young elephant carried by a seasonal river in flood in 2006 with four years of devastating drought a decade later. Yet, life in Turkana has always been a brutal challenge with no easy solutions for improving water security despite the illusion of oil wealth or unlimited groundwater briefly emerging. Lodwar reflects the wider African condition in terms of where demographic growth is likely to increase decisively in the decades to 2050 when the continent will double in size to 2.5 billion people (Department of Economic and Social Affairs, 2022). Increasing water security in small towns is central to the prospects of Africa's development aspirations. The diaries provide insights and lessons into the daily decisions and dilemmas of extremely vulnerable households in this context. We consider three themes from the findings in Lodwar which may chart more sustainable and equitable futures if policy and practice are effectively designed and delivered. One key motif underlines any pathway to a water secure future – an inclusive and safely managed drinking water piped network run by an accountable and efficient utility.

First, the depth and breadth of vulnerability and deprivation in Lodwar is acute and visceral. Beyond the numbers and stories, the lived reality in the town is harsh and unrelenting. Unlike Kitui or Bangladesh, where alternative water resources are available for parts of the year, Lodwar has almost no alternative water sources to the built infrastructure supplied by government and donors. Without a working piped connection to the home, a family's choices decrease as their costs increase. The diaries chart a common story of people using similar volumes of water at around 20 litres per person per day, but very different choices and costs in where and how they access drinking water.

The physical, economic, psychological, and emotional stress of securing basic drinking water from kiosks, rivers, handpumps or neighbours is considerable. Living north of the Kawalase River incurs economic, physical, and water security risks but is a choice some families must take given a lack of alternatives. While families will commonly go a day without eating, few can choose not to collect water given the extreme heat. Without an affordable drinking water supply, this means scooping water from riverbeds in the dry season and taking extreme risks in the unpredictable wet season. Even those living in the main town have to contend with kiosk attendants independently increasing prices to cover their salaries. In cases where ATM kiosks exist, creating a monopoly on the tokens to access the water allows an artificial and inflated market to emerge. This is all predictable in a town where water is extremely scarce, deprivation is high, unemployment is rife, and governance is weak.

Second, decentralisation does offer the prospect for positive change. The county government has undoubtedly inherited a difficult and dysfunctional legacy dating back to the colonial period. LOWASCO is a caricature of a badly managed utility with some of the worst performance metrics across all of Kenya's urban water utilities with an 'expired' tariff and operating licence, and an estimated provision of 24 litres per person per day to less than half of the town's population (WASREB, 2022). A kaleidoscope of individual and uncoordinated donor projects has done little to address the fundamental lack capacity or leadership in the utility despite significant financial investments. Recent legal and policy initiatives scratch the surface of systematically delivering accountability and responsibility into a functioning utility for the town. Development actors share this burden with local offices and highly qualified staff working in partnership with the county government. The results, to date, are inconsequential for the lives of the most vulnerable. Narratives of how ATMs are improving livelihoods ring hollow when the daily practices and behaviour of people are assessed. A new borehole which loses two-fifths of its supply is a poor return.

Dysfunctional systems for a critical resource create market opportunities. While piped systems and kiosks may often run dry, a vending market flourishes with tankers (bowsers) ensuring key facilities (e.g. hotels, government offices, companies) are supplied with water from government boreholes at increased costs. A secondary market in water supply emerges with some households in more distant locations able to invest in storage to benefit from their neighbour's distress. Vending is supplied through county boreholes providing a regular source of income for the government. There is pragmatism in bulk water supply compared to chasing individual connections with complaints from dissatisfied customers. Unfortunately, the spiral of performance heads in the wrong direction. Vended water is by no means an inappropriate or inequitable allocation of public water, but it needs an accountable regulatory framework within a wider strategic planning and financial model. Such a model has been developed with the county government, but it has not been executed (Masinde et al., 2021), though other agencies have endorsed and republished the approach.

Third, Lodwar is at the mercy of the climate and upstream water users. The climate story is well-rehearsed with the critical need to monitor and manage the Lodwar Alluvial Aquifer System. Led by the University of Nairobi, significant progress with county and national government has provided a sound framework for implementation (Olago and Tanui, 2023). If implemented, the town's future looks more sustainable; time will tell on the commitment to these decisions which will limit or stop land sales in key recharge areas where prices have escalated. Of equal concern is plans to increase irrigated agriculture upstream near the Turkwel Gorge dam.

Modelling analysis of scenarios to increase irrigation near the dam reveal this may be the existential threat to Lodwar's water security. If proposed irrigation occurs it will dramatically reduce streamflow reaching the town and recharging the aquifer (Hirpa et al., 2018). County and national government are aware of this threat with the Kerio Valley Development Authority the boundary organisation charged with managing the allocation and development of water resources in the Turkwel river system. As Turkana has witnessed the reduction in flows from the Omo River into Lake Turkana as Ethiopia shifted from using their upstream dams only for hydroelectric to also expanding sugarcane and other crops, the same fate may come to pass if irrigation expands upstream with serious implications for Lodwar's future water security.

6

Conclusion

6.1 Living with the Global Water Crisis

From the hot and humid deltaic plains of Bangladesh to the parched drylands of Kenya, the narratives of water risks are shaped by the drops of rain, the flows of rivers, and the stores of groundwater. In March–April, as rising temperatures, drying ponds, and declining groundwater tables intensify the salinity crisis in Khulna, the first drops of rains bring life to the dry riverbeds of Kitui, filling up earthdams and rock catchments. In July–August, as the tropical monsoon dilutes the pollution in Dhaka's rivers, men and boys dive into the deceptively cleaner waters for respite from the afternoon sun. Meanwhile, women and girls in Lodwar scoop out water from the Kawalase River and roll the heavy jerrycans back home. In October–November, as the second rainy season starts, the fear of flash floods grips the riverbank residents of Lodwar, while the Bay of Bengal becomes a breeding ground for cyclones. As the days and months of the year go by, the nature and distribution of water risks change across geographies. The water diaries of urban and rural Bangladeshis and Kenyans in this book document how people of different cultures in diverse environments cope with these risks while living in poverty.

Despite the stark contrast in sociopolitical history and hydroclimatic contexts, we observed many commonalities in behaviour across our four study sites. Rains stand out as the most defining driver of water source choice, as rural populations in both Khulna and Kitui shift to rainwater, whether harvested in containers from own roof catchments or in rocks and dams in slopes and valleys. Whether in sarees or sarongs, in *kolshis* stacked on the waist or jerrycans balanced on the head, women are the primary drawers and haulers of water. When water needs to be transported via motorcycles or boats, a well is dug or a community tube well is installed, men come into the scene. Behavioural clusters emerge within the same localities based on proximity to infrastructure. Some resort to free and often unimproved sources, while others can afford to pay for vended water from distant sources all year round.

While water expenditures vary between these clusters, food expenditures remain generally stable except for festival times such as Christmas or Eid. Smell, colour and taste of water are stronger drivers of choice, compared to knowledge on invisible pathogenic or chemical contamination without immediate health impacts.

Geography and culture also shape differences in acceptable level of risks, behaviours, and policy outcomes. While handpumps were introduced in Kenya and Bangladesh at roughly the same time, high population density and easy availability of groundwater at shallow depths spurred the growth in local markets and private tube wells in rural areas like Khulna, whereas in Kenya and much of Sub-Saharan Africa, handpump installation continues to be a donor-funded investment. In Khulna, while drinking water access may be constrained by salinity, there is an abundance of water for washing, bathing and livestock. In Kitui and Lodwar, where limited water availability contributes to daily collection of around 20 litres per person per day, every drop counts. Yet, our diary respondents often reported having sufficient water as their habitus of place, culture and practice has influenced what is 'acceptable'. For Dhaka's marginalised people, bathing and fishing in the polluted rivers might be acceptable, though the city's middle and upper class residents would maintain their distance from these waterbodies. For our Khulna diary participants, the photo of scooping out water from dry riverbeds in Kitui felt alien. Likewise, with the photo of the Khulna boat vendor transporting water through water, our Kitui participants found it difficult to conceive the notion of water scarcity in Khulna.

Individual practices are linked to institutional behaviours, embodying the legacies of colonial and post-colonial aid-dependent policy regimes. While a series of policies outline the roles of public and private actors, from national to local levels, these often remain on paper with limited impact in practice. Regulation is missing or ineffective for rural drinking water services in Kitui and Khulna, while non-compliance is normalised in case of urban water pollution in Dhaka and unreliable piped supply in Lodwar. In Dhaka, private governance by global fashion brands and consortiums, and mounting civil society pressures have emerged in response to weak regulatory enforcement of river health by the state. In Khulna and Kitui, absence of effective monitoring and coordination have resulted in overlapping investments in rural water infrastructure by donors and governments and sub-optimal usage of financial resources. Regardless of technology type and user capacity, infrastructure is routinely handed over to communities assuming that since they need it, they will maintain it. Extensive private investments by households and small enterprises are a response to the poor sustainability of public investments, posing uncertain water quality risks and high-cost burdens. The proliferation of around 16 million shallow tube wells in rural Bangladesh (Fischer et al., 2020), for instance, may have facilitated the MDG target of increasing access, but it now creates an immense behavioural challenge to convince people to shift away from their own sources and pay for publicly provided higher service levels such as piped schemes. The process to influence local habits and cultural norms will take time, patience and money.

6.2 Responding to the Global Water Crisis

So, what lessons can we take to guide policy and practice to do better in the future? Two areas deserve attention – first, the interactions between rainfall and water use behaviours and the implications for public health and financial sustainability of waterpoints in a changing climate; and second, the need for better information on water risks for institutional accountability and sustainable finance for delivering impacts at scale.

As rains replenish wells, ponds, and storage tanks, and washes away pollutants from rivers and floodplains, people's water source choices and river use behaviour take a sharp turn. For households with low and variable incomes, use of rainwater reduces the collection burden and expenditures associated with alternative sources. Besides the rationale of cost and convenience, these behavioural dynamics are also driven by deeply ingrained psychological and cultural dispositions. Paul et al. (2018)'s study in Ethiopia found trust in social institutions as a predicator of choice, as the need for community level coordination and management varies across source types – being highest for boreholes and lowest for surface water sources. Borehole use was associated with lower-income, and users exhibited lower trust, worrying about competing water access and others not doing their share of maintenance. The freshly collected rainwater and rising river flows evoke a sense of abundance and natural blessing amongst the rural residents of Khulna and Kitui and urban riverbank populations in Dhaka, whose lives and livelihoods are intricately linked to water. However, this rainfall driven behaviours have two major implications for water policy – public health and financial sustainability of rural water services.

Water quality results from Kitui and Khulna confirm the high prevalence of faecal contamination in ponds sand filters (Figure 3.7), earth dams and rainwater storage tanks (Figure 4.4), with increased risks in the wet season even among shallow groundwater sources. This is supported by various studies showing that in contexts with multiple sources, households may not choose the safest option (Hamilton et al., 2019, Foster and Willetts, 2018), particularly when the contaminant does not affect taste, smell or colour, as in the cases of arsenic and fluoride as well. Those living in poverty, may choose to discount a possible future health benefit over an immediate water payment cost (Ray and Smith, 2021). In Dhaka, men, women and children are daily exposed to high-levels of pathogens while swimming and bathing in the rivers in monsoon. Modifying behaviour for risks that do not have immediate consequences is challenging and complicated by the lack of monitoring and risk communication.

Kiosks and handpumps in Kenya suffer from a significant drop in revenues in the wet season, to the extent that many piped schemes in Kitui close operations seasonally (see Section 4.5). This is more prevalent for pay-as-you-fetch services, as also observed in the case of small water enterprises in coastal Bangladesh. Intra-seasonal variability is shown to moderate these revenue fluctuations, as users will value continued operations of piped schemes and handpumps if there is likelihood of dry spells interrupting wet seasons. This may be a potential silver lining in influencing use of improved sources during the wet season, given the increased trend of sub-seasonal rainfall variations in certain regions of Africa (Dyer and Washington, 2021). Climate shocks such as cyclones, floods and droughts also disrupt functioning of water supply systems posing higher cost burdens on users. We saw this in the cases of LOWASCO's boreholes being flooded (see Section 5.4) and Khulna's pond sand filters being filled with saline storm surge water during cyclone Amphan (see Section 3.4).

A key response to the water crisis is to improve institutional accountability and decision-making through timely and accurate data on risks. Global and nationally representative aggregate statistics are useful for public advocacy and sector funding, but not for defining appropriate responses to locally specific challenges. In Dhaka, our river water quality monitoring data have helped calibrate hydrodynamic models to simulate impacts of different policy interventions such as river flow augmentation, planned construction of sewage treatment plants, and relocation of tanneries (Bussi et al., 2023, Whitehead et al., 2019, Whitehead et al., 2018). The river diaries have illustrated the social inequalities in pollution exposure across space and seasons, supporting the call for risk communication and provision of better water and sanitation facilities to reduce exposure in the short term. Installation of automated water quality monitors can improve institutional accountability by identifying pollutant discharge events, address the information gaps for tracking progress towards SDG 6.3, and communicate real time risks to river users.

In case of drinking water services, we have shown how data on 'one main source' can be misleading in rural contexts with multiple source use. While water diaries may not be systematically implemented at large scales due to logistical and budgetary constraints, traditional household surveys can be modified to incorporate issues of multiple sources, quality, costs, and water insecurity experiences, potentially on a sub-sample of the survey population. The inclusion of arsenic and *E. coli* analysis in the 2019 Multiple Indicator Cluster Survey Bangladesh, for example, shows the value of better data in challenging traditional statistics on access (see Section 3.3). Performance of water supply infrastructure remains a grey area for rural service delivery, as most developing countries like Bangladesh and Kenya lack comprehensive inventories of public and donor-funded waterpoints, their functionality, and water safety. Professional service delivery models emerging in Asia and Africa can plug these gaps through regular reporting on performance metrics to attract results-based funding (McNicholl and Hope, 2024). Such data would be consistent with the diary data but be monitored as service delivery contracts providing governments and donors with more granular, longitudinal and effective information.

As described in Section 3.5, the Government of Bangladesh has collaborated closely with research aligned to the diary work to pilot a results-based contract for safe drinking water in the coastal zone. The SafePani model will guarantee safe drinking water for 1,200 schools and healthcare facilities from 2024 to 2030 aligning with the government's commitment to SDG 6.1. A central feature of the model is co-funding of the operational costs with the government committing 45 per cent of the operational costs for seven years, which will be matched by external donors. In contrast, current funding for drinking water is bundled into a wider set of facility management service costs (energy, sanitation, buildings, etc.) with no effective monitoring or evaluation of results. At a cost of less than USD 1 per person per year, the SafePani costs are modest whilst the outcomes for health and education are likely to be high. Quarterly data on water safety (faecal and chemical contamination, sanitary inspections) and service reliability (breakdown events and response times), reported by the professional service provider, are being verified by third party checks and sensor technologies.

The critical importance of ensuring functioning waterpoints in dry periods, as revealed by the rainfall driven behavioural shifts documented by the water diaries, has contributed to funding support for professional water service delivery. In Kenya, FundiFix has progressively increased its share of funding from corporate partners with performance on water safety being incorporated as new metrics for results-based funding. Generous support from corporations and charitable foundations also catalysed a global results-based funding initiative led by Uptime with 5 million rural users in 16 countries as of 2024 (McNicholl and Hope, 2024). Sustaining and scaling out these models require high-level political support and donor cooperation, given that cost recovery from users is challenging. While Kenya's Water Act 2016 expanded the scope of service providers from community management and formal utilities to include private actors, the government is yet to act on these opportunities. The SafePani model provides one example of government leadership and action with financial commitment, though it is rare.

Water supplies in schools and clinics lie under government mandates. There is less local political interference compared to communities where powerful individuals control the local revenue and have no interest or incentive to lose this source of money. The process of scaling up community water supplies is fraught with vested interests and political obstacles which stymie, slow and generally stop progress. Benin[1] and India[2] are two examples of countries embarking on ambitious national models to systematically address the unsatisfactory nature of the current situation with the financial support of the World Bank (World Bank, 2018), in the former, and the leadership of the Prime Minister (Ministry of Jalshakti, 2024), in the latter. If successful, these could serve as important and influential national examples for wider policy adoption and adaptation in the future.

6.3 Investing in Partnerships for Sustainable Results

Our final reflections consider the role of partnerships for progressive achievement of water security in line with the SDGs. There is a saying that *'science without policy is science, and policy without science is gambling'*. Interdisciplinary collaboration with governments and practitioners is often touted as the means to identify and remove such barriers to progress. There are useful lessons for the design and funding of international development programmes based on our research and impact-oriented work under the REACH Programme.

REACH's long-term funding commitment, extending from £15 million for seven years to £22 million for 10 years, allowed us to implement a suite of innovative social and biophysical research in each of our study sites or 'observatories', aligned to the interests and priorities of government partners. The diary work emerged in the early days of the REACH Programme as it became clear that any meaningful interdisciplinary work required greater efforts to align biophysical data with social data. While climate scientists can generate 15-minute data points of rainfall, temperature and humidity at regional scales, social researchers often struggle to generate longitudinal evidence on local practices. Designing and implementing year-long diaries require clear understanding of the study populations and contexts, which could be generated from household surveys, infrastructure mapping or water quality monitoring, which in turn, take years to develop, implement and analyse. The behavioural dynamics revealed by the diary study also prompted the need to advance our understanding of rainfall events, such as the risks of intra-seasonal variations or delayed onsets. Like the diary work, we and our colleagues also engaged in other longitudinal research, such as the year-long water quality monitoring in Kitui, the year-long recording of operation and maintenance expenditures and user payments by water point managers in Khulna,

---

[1] The Rural Water Supply Universal Access Program Project for Benin aims to increase access to water supply services by supporting investments in piped rural water supply and strengthen operational and financial sustainability by delegating service delivery to private operators through regional performance-based contracts, supported by appropriate tariff policy and regulation arrangements.

[2] The Jal Jeevan Mission aims to provide safe and adequate drinking water to all households in rural India through individual household tap connections by 2024. It is based on a community approach involving extensive information and education campaigns.

and the monthly analysis of river water quality in Dhaka (refer to Section A.4 in Appendix). These research activities complemented one another and helped explain the distribution of risks and drivers of behaviours, providing new insights for driving action. This would not have been achievable with short-term funding.

Research activities often end with a portfolio of recommendations for policy-makers, who usually are never engaged in the process and unaware of the findings published in academic journals. Driving research into action requires long-term engagement with practitioners through regular in-person meetings, workshops, and co-authored publications. These relationships are hardly restricted to one project and span over years of collaboration through multiple work streams. UNICEF has been a core global, regional and national partner of REACH, and the motivation of key staff has been instrumental in facilitating research impact. In Bangladesh, UNICEF's longstanding technical support to the country's water, sanitation and hygiene sector created opportunities to engage with key government agencies, including DPHE, the Policy Support Branch of the Local Government Division, the Directorate of Primary Education, the Directorate of Secondary and Higher Education, and the Deputy Commissioner of Khulna district. These networks paved the way to co-designing, piloting and scaling up the SafePani model for schools and healthcare facilities and mobilise government funding. The water diary work gained traction among UNICEF staff in Ethiopia, leading to the method being piloted to evaluate impacts of water, sanitation and hygiene interventions in refugee camps. The methodology and findings also informed a global initiative to monitor drinking water affordability, led by the UNICEF Headquarters in New York (see WHO/UNICEF (2020)). The diaries challenged affordability norms derived partly from practices in the global north where there is fairly standard use of piped water in the home on a daily basis.

The river diaries in Dhaka prompted policy discussions on distributional implications and investment priorities, through our colleagues at BUET who have been closely involved with the government and donor organisations from before the inception of REACH. Water quality planning and management for Greater Dhaka is based on fragmented, ad hoc, and suboptimal data that do not address the watershed dynamics at a system level. The REACH Programme's water quality monitoring data, as presented in Section 2.3, supported modelling exercises to evaluate the potential impacts of the 12 proposed sewage treatment plants to be constructed in the next two decades under the DWASA's sewerage masterplan. In line with the river diaries, which revealed highest vulnerability for populations along the Tongi Khal, the modelling study concluded that the construction of the Tongi Khal sewerage treatment plant needs to be brought forward to achieve the greatest and earliest benefit for exposed populations (Bussi et al., 2023). Based on recommendations by the Bangladesh Water Multi-Stakeholder Partnership, which partly draws on REACH's river water quality monitoring, the World Bank funded 'Bangladesh Environmental Sustainability and Transformation' project will establish the first DoE network of 22 continuous surface water quality monitoring stations to monitor real time water quality of Dhaka rivers (World Bank, 2022).

The positive direction of the improvements discussed have partly been achieved through navigating the regular rotation of government and donor staff, who hold positions of influence in driving, slowing or ignoring change. Our long-term engagement with many local partners has provided political capital and trust to transition the work and maintain momentum, despite rotation to individual staff and disruption caused by COVID-19 pandemic. Some individuals with vested interests in the status quo have actively campaigned against change. In case of rural water services, the shift to investing more in low-cost maintenance of waterpoints is a threat to traditional high-cost capital investment programmes, where contracts and payments benefit many people in a short period. The change also questions those agencies with existing responsibilities for constructing high-quality waterpoints with good water safety. The availability of higher-quality information on actual water practices challenges assumptions on the effectiveness of current practices. This is not welcome by some as careers and promotion depend on perceived progress of current work. As most research engagements usually conclude quickly, influential local actors have time on their side to challenge or ignore inconvenient new evidence.

The water diaries have presented daily practices to water insecurity risks in Bangladesh and Kenya from the lives of those most at risk. Those risks are predicted to increase over time with uncertainty in where and when they will occur. International climate finance is increasing in response to these significant but uncertain local adaptation and global mitigation challenges and costs. Whether this finance will reach the most vulnerable is unclear. While academic terminology and policy jargon can seem to endlessly produce new terms and frameworks, there are simple measures of improving water security for the poor. Improving river health and delivering safely managed drinking water are universally compelling though beguilingly difficult. The river never lies, unlike vested interests from those gaining by over-abstracting water or avoiding treatment costs, so effectively monitoring river health is a necessary condition for all actors to focus action on outcomes which benefit rivers, society and economies in the short and long term. Safe drinking water services are equally a common and critical measure of government performance. If a government cannot achieve and sustain this fundamental anchor of development and prosperity, it is hard to imagine how associated goals for education, health and prosperity will be sustainably achieved. Water diaries offer a modest lens to examine global policy and local practice challenges to view the conditions and the choices vulnerable people contend with each day in search of a more sustainable and equitable future.

Appendix

Making of the Water Diaries

The narratives of water risks, practices, and institutional responses we present in this book draw on seven years of extensive research by the REACH team. Funded by the United Kingdom Commonwealth and Foreign Development Office (UK FCDO), the REACH Programme (2015–2024) was designed to improve water security for 10 million poor people in Asia and Africa by mobilising science to support governments and practitioners in making policy and investment decisions. The REACH Programme adopted a risk-based approach for framing the global water crisis, acknowledging the multiple natural and 'manufactured' drivers of society's water challenges. A risk framing motivates interdisciplinary research to capture the trade-offs across competing values and objectives across scales and reduce uncertainty in decision-making.

Building on this risk-based approach, we designed our water diaries as a lens to understand how people navigate multiple water related risks in their daily lives. For Khulna, Kitui, and Lodwar, the diaries focused on household drinking and domestic water choices (referred to as 'Water Diaries', Section A.2), while in Dhaka, it focused on people's interactions with river water (referred to as 'River Diaries', Section A.3). The analysis of the diaries informed and was informed by a portfolio of social and biophysical research (Section A.4), led by us and our colleagues across the REACH Programme, based at the University of Oxford, BUET, Dhaka University, the International Centre for Diarrheal Disease Research (icddr,b),UNICEF-Bangladesh, and the University of Nairobi. The survey of schools and water supply infrastructure in Kitui county, outlined in Sections A.4 and A.5, were funded by the USAID Sustainable WASH Systems in collaboration with the REACH Programme. The household survey and water infrastructure mapping in Polder 23 of Khulna district were funded by the Research England Internal Global Research Challenges Fund.

Ethical permission for all research activities was obtained from the University of Oxford's Research and Ethics Guidelines to ensure informed consent, confidentiality, and no harm to all participants. In Kenya, REACH research was registered with the Government of Kenya's National Commission for Science, Technology and Innovation.

A.1 Diaries: Motivation and Previous Work

The inception of the REACH Programme coincided with the beginning of the SDG era in 2015, with the term 'water security' gaining increased prominence among academics and practitioners. Alongside the political urgency to devise plans and chart pathways to reach the SDG targets by 2030, the need for new metrics and methods became more apparent. One such metric was 'affordability' – which is not only about how much people spend on water, but what they give up to access water services and what it means for people who prefer to use unpaid or unimproved sources. This seasonal shift to surface water sources proved to be particularly difficult to sustain reliable water services in rural Kenya, where the University of Oxford has incubated a professional maintenance service delivery model (FundiFix) since 2013. Understanding the nature and drivers of water use behaviour – the choices people make in their daily lives across different contexts – could better align policy and practice to cultural values and preferences. By one definition, 'water security' is an 'acceptable' level of water-related risks, yet what is 'acceptable' for individuals and communities was a knowledge gap implicitly embedded in their daily choices.

We were inspired to chart these everyday behaviours through novel methods that went beyond the 'reductionist' approaches of large-scale surveys that traditionally fed the aggregate statistics of water access, without acknowledging the risks to safety, reliability, affordability, and equity of water services emerging from use of multiple sources. We drew inspiration from Collins et al. (2009)'s 'Portfolios of the Poor', which developed the 'financial diaries' methodology to study the financial practices of households living in villages and urban slums of South Africa, Bangladesh, and India. Diaries have been used extensively in psychological and health research (e.g. Wiseman et al., 2005, Cates et al., 2004, Lawson et al., 2004); however, there have been limited examples of its application in studying water use behaviour (e.g. Bishop, 2015, Harriden, 2013, Wutich, 2006) prior to our work.

The diary method is an instrument for individuals/households to record changes in daily processes or practices which may be subject to unpredictable shifts in behaviour or outcomes, for example, the effects of seasonality on household incomes and expenditures. In such cases, simple 'snapshots' of behaviour at a particular time may not capture the temporal variations. Compared to other research tools, diaries are less likely to suffer from problems of recall bias as they rely on short-term memory (Bolger et al., 2003). Wutich (2009) found that the diary method yielded the most accurate estimate of per capita water use over a week compared to prompted recall and free recall methods, which either underestimated overall water use or missed out relatively low-volume water use tasks like washing and cleaning. However, as diaries are produced by participants in their own time and setting in absence of the researcher, participants need to be trained thoroughly to ensure accuracy of data being recorded and minimise confusions in making entries (Wiseman et al., 2005). Regular communication between the researcher and the participant is required to keep the latter motivated and build trust between both parties. This can restrict the sample size due to resource constraints, creating a trade-off between breadth and depth of data collected.

The design and implementation of the diary method is often guided by issues relating to (1) the structure and content; (2) duration and frequency; (3) respondent attrition and fatigue; (4) compensation; and (5) use of complementary methods. Water diaries intended to capture household water use behaviour usually involve structured charts, outlining the sources, purposes, and volumes of water used by individuals (e.g. Harriden, 2013, Wutich, 2006). However, if the research requires participants to record the social interactions embedded in their daily quest to access to water and reflect on these events from their own perspectives, the researcher may design an unstructured or semi-structured diary (Bishop, 2015). As diaries usually require participants to read and write or have someone to make entries on their behalf, pictorial diaries often proved to be more appropriate in settings with high levels of illiteracy. Wutich (2006), for example, used illustrations of different water sources, water use tasks, and container types to estimate the source and volume of water used by each household member for consumptive, hygiene, and domestic needs in an urban slum in Bolivia. While pictorial diaries can potentially overcome the literacy barrier, care must be taken to ensure that illustrations are sensitive to cultural perceptions (Wiseman et al., 2005).

The duration and frequency of the diary keeping exercise largely depends on the data requirements of the research. Shorter diaries, maintained over a few days to a week, require less time commitment from the participants and are unlikely to be affected by fatigue or dropouts. Harriden (2013)'s study of intra-household water use behaviour in Australia, for example, required participants to record all water use activities over a week, particularly noting who used water, for how long, in what quantity, at which time, and for what purpose. Longer diaries, on the other hand, can suffer from respondent attrition and research fatigue, but may be necessary to capture temporal variations. A noteworthy example is Wiseman et al. (2005)'s study of financial transactions in rural Tanzania and the Gambia, where participants were asked to maintain a pictorial financial diary every day for a year. The authors noted a dropout rate of around 20 per cent and found that successful maintenance of longer diaries depended on the level of trust between the diarist and the field researchers, who visited the diarists regularly to keep them engaged. It is important not only to note the dropout rate but also ensure that those who dropped out are not systematically different from the whole population. Longer diaries can also create a 'conditioning effect', whereby participants may become tired of keeping records on similar-seeming activities leading to abbreviated or less thorough entries (Wiseman et al., 2005). If they miss an entry, they may also go back and 'fill in' what they missed, thus, undermining one of the core purposes of using diaries (Bishop, 2015, Bolger et al., 2003).

Since diaries require long-term commitment from the participants, researchers often provide financial incentives to motivate participants or to compensate for their time and effort. This raises methodological and ethical concerns among the research community. As experienced by Meth (2003), offering payments for participation can specifically attract economically vulnerable people and may cause resentment among those not selected for the study. Others argue that the need for compensation depends on the complexity of task required (Bartlett and Milligan, 2015). The water use behaviour study by Wutich (2006), where each household was offered USD 2.50, involved day-long diary keeping by each household member, followed by extensive interviews that required participants to recall their water use activities during the preceding week.

Diaries are often combined with alternative research tools such as interviews, observations, questionnaire surveys, and focus group discussions (FSD Kenya, 2014, Wutich, 2009, Wiseman et al., 2005). These are necessary for collecting baseline data that can better inform the diary design, for engaging participants at different stages of the research process, for ensuring compliance and proper recording of events/activities, for keeping up participants' morale, and most importantly, for triangulating data from different modes of enquiry. For example, the Collins et al. (2009)'s 'financial diaries' methodology involved baseline questionnaire surveys on demographics, income sources, assets, and financial tools, followed by year-long bi-monthly financial diary visits during which interviewers captured detailed data on all cash flows over the preceding two weeks, as well as any events that may have influence household welfare during that period.

A.2　Water Diaries: Household Water Source Choices in Khulna, Kitui, and Lodwar

A.2.1　Designing the Diaries

Our diary methodology was aimed to capture households' water source choices, which can be shaped by a range of concurrent factors, including rainfall variability, operational disruption of infrastructure, costs of water, household income and expenditures, and time spent in collecting water. Given the low literacy level among rural participants, we made the diary as simple as possible so that it was easily comprehensible by the respondents, and not considered too burdensome to fill in on a daily basis. Pictorial charts were used to record data with written input from participants being limited to ticks, crosses, and numbers. Each day's diary was restricted to two pages to minimise printing costs and reduce paper use. The charts comprised of three sections – (1) water sources, amount of water fetched, cost of water whether payments were made or not, and collection responsibilities disaggregated by gender; (2) sufficiency for drinking, washing, bathing, livestock, and small-scale irrigation (only for Kenya); and (3) household expenditures on food, education, health, transport, energy, and miscellaneous items.

The preliminary diary design was pretested in March 2017 with adult women from an all-female water user committee in Kyuso town in Mwingi-North sub-county. The design was based on an extensive review of the literature on the diary method and the state of the water supply situation in rural Kenya, as well as the context specific knowledge and expertise of researchers working in the region. We invited about 15 women to attend a 2-hour focus group discussion, of whom 11 attended. Women were intentionally recruited as they are usually responsible for fetching water for the household and hence, have the best knowledge on this matter. The purpose of the focus group discussion was to explain the diary method to the participants and identify whether the methodological design was appropriate for the local context and easily comprehensible by the participants. Following the group training, the participants were given printed copies of the diary sheets and requested to fill them every day for the next month. During this month, the research assistant visited all households every week to ensure they were filling the diaries accurately, and also enquire about noteworthy entries, including unusual expenditures, change in water sources, or high volumes of water collected. Analysis of this pilot diary data, elaborated in Hoque and Hope (2018), showed promising results indicating the potential for scale up to a large sample size over a longer time period.

The diary design for the main study varied slightly across the study sites to account for contextual differences. Unlike the Kitui water diaries shown in Figure 1.2, the Khulna diaries did not have the section for sufficiency, as quantity of water for domestic purposes is not a concern in Bangladesh (Figure A.1). The quantity of water used per day could be easily quantified in Kenyan contexts, as households generally used the same source for all purposes in a given day, water was mostly collected in standard sized jerrycans, and use of water at source, for example, washing and bathing at a pond, was rare. However, calculations on consumption per capita are aggregate estimates, as part of the water may have been used for livestock or house construction purposes. For Khulna, the amount of water could only be quantified when fetched from off-site sources such as vended water or

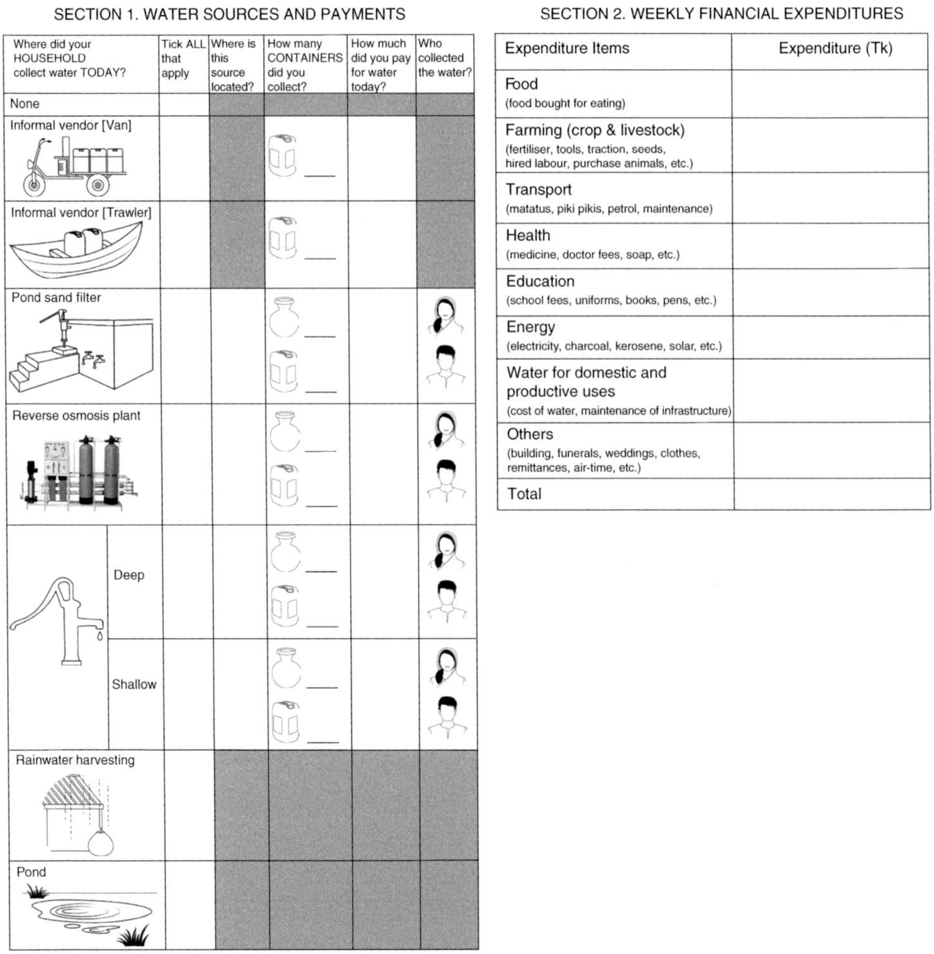

SECTION 1. WATER SOURCES AND PAYMENTS

Where did your HOUSEHOLD collect water TODAY?	Tick ALL that apply	Where is this source located?	How many CONTAINERS did you collect?	How much did you pay for water today?	Who collected the water?
None					
Informal vendor [Van]					
Informal vendor [Trawler]					
Pond sand filter					
Reverse osmosis plant					
Deep					
Shallow					
Rainwater harvesting					
Pond					

SECTION 2. WEEKLY FINANCIAL EXPENDITURES

Expenditure Items	Expenditure (Tk)
Food (food bought for eating)	
Farming (crop & livestock) (fertiliser, tools, traction, seeds, hired labour, purchase animals, etc.)	
Transport (matatus, piki pikis, petrol, maintenance)	
Health (medicine, doctor fees, soap, etc.)	
Education (school fees, uniforms, books, pens, etc.)	
Energy (electricity, charcoal, kerosene, solar, etc.)	
Water for domestic and productive uses (cost of water, maintenance of infrastructure)	
Others (building, funerals, weddings, clothes, remittances, air-time, etc.)	
Total	

Figure A.1 The water diary charts designed for Khulna, Bangladesh, which were translated to Bangla. One hundred and twenty households from the southern part of Polder 29 participated in the diary study from May 2018 to April 2019. Reprinted from Hope and Hoque (2020) under the terms of the CC BY 4.0 license.

pond sand filter, but not for sources on-site sources such as tube wells and rainwater harvesting. Hence, while we could estimate 'amount of water in litres by source' fetched by the four expenditure clusters in Kitui (Figure 4.5), we used 'proportion of days per month' a given source was used by the four expenditure clusters in Khulna (Figure 3.6). For Kitui and Lodwar, the column on 'Who collected the water?' included boys and girls as well; however, in Khulna, although children accompanied mothers for water collection, their involvement was minimal, and hence, were excluded from the diary chart.

The need for simplicity inevitably restricted the depth of data collected. Firstly, while households ticked the type of water infrastructure used, details of the specific waterpoint were not recorded, making it difficult to match it with the water audit data. In Kitui and Lodwar, for example, participants continued to tick hand-pump, even when they shifted from one handpump to another due to breakdown or fall in water table. In Khulna, there was an additional column on 'Where is the source located?', but entries were often ambiguous like 'neighbour' or 'market'. Secondly, we did not ask respondents whether the water was used for drinking or other purposes, which was particularly problematic in Khulna where different sources were often used on the same day. Third, we did not track changes in the number of people living in the house each week, or the livestock herd composition and size, which would have affected water use. This was particularly important for Kenya, where children often stayed away from home on weekdays or school term time, and many adults lived elsewhere from time to time for work.

A.2.2 Sampling and Training Diary Participants

The household survey datasets (see Section A.5) provided the sampling frame for the water diaries in Khulna, Kitui, and Lodwar. A subset of the surveyed areas was selected for the diary study – southern part of Polder 29 in Khulna, Kyuso and Tseikuru wards of Mwingi-North subcounty in Kitui and within the piped network area in Lodwar town. Survey data on 'wealth status' (poor or non-poor) and 'concerns for water' ('water is costly', 'water is unsafe to drink' and 'others') were used as selection criteria (see Section A.5). 'Water is costly' and 'Water is unsafe to drink' were considered as proxy indicators of affordability and quality respectively. A 3×2 matrix was created, allocating 25 households to each of the four main categories (Poor + costly, Poor + unsafe, Non-poor + costly, Non-poor + unsafe) and 10 households to each of the two other categories. An additional 30 households, five from each category, were selected for the backup list.

The sampling strategy involved three steps: First, the data was cleaned by removing households without at least one contact number and missing geolocations. Second, the household list by disaggregated by 'diary category', resulting in six lists and shapefiles. Third, ArcGIS 10.5 was used to randomly select 150 households from the sampling framework, 30 from each of the first four diary categories and 15 from each of the last two. The additional 30 households, five under each category, were part of a backup list used to recruit households in case the ones from the primary list refused to participate. In Khulna, given the relatively lower costs and more manpower in the local team, all selected households were visited in-person, and both husband and wife were invited for a 2-hour training session at the REACH Polder 29 office, with transport prearranged by the team. The trainings were led by the author (Sonia Hoque) and held in Bangla over six sessions March 2018, with groups of 20 households and approximately 40 people in each. In Kitui and Lodwar, the diary trainings were held in May 2018 and led by research assistants in the Kikamba and Swahili (and Turkana language), respectively. Households were contacted by phone and one adult member was invited from each selected household. If any of the preselected households were unavailable or unwilling to attend the training, another household was selected from the backup list. Participants were compensated with cash for their time and transport expenses.

The trained households were asked to complete a one-week pilot study which gave them an opportunity to better understand and practice filling the diaries. During the week, the research assistants visited all households in turn to ensure that participants completed the tasks accurately. More importantly, it allowed the research assistants to train other household members, such as school going children, where the adult participant struggled with literacy. It was vital for the research assistants to build a strong rapport with their allocated households – a relationship that kept participants motivated over the course of the study and appreciate the value of their engagement with research. At the end of the week, the research assistants entered the diary data in a digital form programmed in ONA (a mobile data collection platform) for it to be checked by the lead researcher.

A.2.3 A Year of Diary Keeping

The main phase started a week after the pilot phase and continued for 52 weeks – from 28 April 2018 in Khulna, and 31 July 2018 in Kitui and Lodwar. Though we invited 120 households for training in each site, the main phase in Kitui and Lodwar started off with 115 and 98 households, respectively, owing to non-attendance and dropouts post pilot. In Khulna, a couple of households who dropped out post pilot were replaced with others from the backup list; hence, the main phase started off with 120 households. During the main study, research assistants visited the households bi-weekly to collect the completed diaries (Figure A.2) and ask follow-up questions on any noteworthy events such as health problems, visits by family and friends, large purchases or water infrastructure breakdown that may have affected their water use and spending behaviour. The data collected was submitted bi-weekly through ONA and checked by the lead author regularly so that data-entry errors and queries can be addressed promptly. There were two enumerators in Khulna, each responsible for 60 households, while Kitui and Lodwar had four enumerators each.

Retention of participants and regular submission of data seemed particularly difficult for Lodwar, as reflected by the weekly diary submission data (Figure A.3). Lodwar's harsh climate, difficult transport routes, the high mobility of the population and the need to visit the same household multiple times, made it challenging for enumerators to maintain close contact with participants as often as planned. In

Figure A.2 Water diary sheets filed by household ID and stacked in REACH Polder 29 office in Khulna.

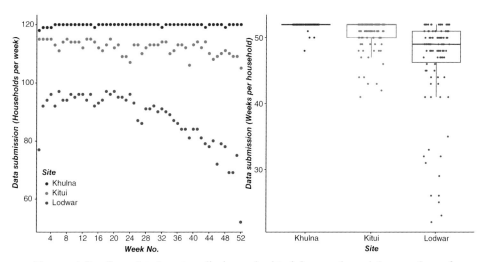

Figure A.3 Completed water diaries submitted by week and by number of households in Khulna, Kitui, and Lodwar.

Kitui, the large distances between households meant that enumerators could not visit all households at the same interval but maintained regular contact with them over phone. The dedication and intelligence of enumerators mattered greatly in keeping participants motivated, recognising data errors, and responding to data queries. It was ensured that all households retained for analysis had at least 35 weeks of data.

Compensation was critical to keep households motivated for such long-term engagement. In Khulna, each household was delivered a packet of essential groceries (such as cooking oil, rice, pulses, and soap) worth USD 5 at the end of each month. The selection of items was based on feedback from the participants, ensuring that the needs of women participants were particularly met. In Kitui and Lodwar, participants were given store vouchers worth USD 5, as it was logistically not possible to buy and transport goods every month.

A.2.4 Data Analysis

The water diaries generated daily data on water sources, amount and payments, and week wise data on food, education, healthcare, and other expenses. For analysis, monthly totals were calculated for all required variables, adjusting for missing days. For example, in Khulna, the proportion of days per month that a household used each of the six available water sources (i.e. deep tube well, shallow tube well, pond sand filter, vendor, rainwater harvesting and pond) was calculated, while in Kitui, the total volume of water collected over the month was calculated for each of the nine sources (i.e. earthdams, hand-dug wells, handpumps, kiosks, dry riverbed scooping, vendors, rainwater harvesting, rock catchments, and rivers). Data for missing weeks were substituted with averages for the weeks available for the given month for that household.

For identification of expenditure groups In Kitui and Khulna, a k-means cluster analysis was carried out in IBM SPSS 23 using household monthly water expenditures as the input variable and 4 as the number of clusters. K-means clustering segments the data in such a way that the within-cluster variation is minimised (Mooi and Sarstedt, 2011). Descriptive statistics and plots were used to characterise source choices and expenditure patterns of the four identified clusters, which were named as high regular expenditure, moderate regular expenditure, no/low expenditure, and seasonal expenditure groups, respectively. While there were noticeable differences in the wealth status and location of households between the four clusters, these were not statistically significant. In Lodwar, cluster analysis was not conducted owing to absence of seasonal or spatial patterns in water source choices and expenditures within Lodwar Water and Sanitation Company (LOWASCO)'s service area. Similar findings were observed in our water diary study in Wukro, Ethiopia, a small town with unreliable piped water supply (Grasham et al., 2022).

A.3 River Diaries: Direct Observation of River Use Behaviour in Dhaka

In Dhaka, we designed and implemented a 'structured direct observation' study to monitor the gender- and age-disaggregated daily river water use practices in relation to the spatial and temporal variations in water quality risks. Direct observation is an established method in social science research whereby the researcher uses a pre-designed questionnaire to collect standardised quantitative information on the research subjects in their usual environment without any alterations. Early examples of direct observation can be found in the medical anthropology literature, where researchers observed the patterns of human contact with water bodies to evaluate the pathways of schistosomiasis transmission (Slootweg et al., 1993, Dalton and Pole, 1978, Dalton, 1976) with more recent applications is monitoring handwashing behaviour in rural South Asia (Halder et al., 2010, Ram et al., 2010).

We conducted our study in two phases, inspired from a one-week observation study along Tongi Khal led by Villanueva (2016) as part of his MSc dissertation. The first phase was carried out over an 18-day period in the dry season (9–26 February 2019) which coincided with the Bishwa Ijtema held in two groups of three consecutive days. The second phase was conducted for a 15-day period in the wet season (20 August–3 September 2019). We selected ten observation points, two of which were in Zone-1, seven in Zone-3, and one in Zone-4. The site selection involved multiple scoping visits to the household survey areas to identify spots with observable river use activities, as well as ensuring spatial distribution, diversity of river interactions, accessibility and security for field team, and alignment with river water quality monitoring points. Zone-2 was excluded as we did not find any interactions with the river during our scoping visits, as the short river branch flowing through the bottom part of Zone-2 remains dry for part of the year. During analysis, as shown in Figure 2.5, we merged results from the ten observation points into six sites in three zones, namely, Konabari (Zone-1), Ijtema field, Tongi slum, Abdullahpur and Railway Bridge (Zone-2) and Mausaid (Zone-3).

Each observation day comprised three-hour slots: 7–10 am (morning); 10.30–1.30 pm (midday); and 2.30–5.30 pm (afternoon) during which enumerators recorded their observations in an electronic form in a tablet, whereby an 'observation' is defined as any activity conducted by an individual or group visible within the enumerator's field of view. Thus, each slot comprised several observations, with each observation including one or more activities. Over the 33 study days, we recorded about 7,900 observations for 852 slots. River users were visually categorised as children and adolescents (less than 16 years of age) and adults, as male and female, and also as groups and individuals. Differentiation of age was based on enumerators' judgment; some adolescents and young adults may have been misclassified. Activities were listed as drinking, food washing, water collection, dish washing, laundry, washing oneself, bathing, using hanging latrine, open defecation, urination, fishing, swimming, boating, and other. These activities were listed as multiple-choice questions, as multiple tasks were often conducted by different individuals in a group. Enumerators were given clear guidelines for certain activities; for example, if fishing was done on a boat, the activity was recorded as fishing and not boating. Enumerators also took a photo for each observation, following approved ethical guidelines, which were inspected by the lead author to resolve issues in data entry.

A.4 Diaries as Part of Interdisciplinary Research

The water diaries in each site were complemented by a suite of social and bio-physical research methods depending on the broader research objectives of the REACH Programme. The methods outlined (Figure A.4) are those relevant for the data presented in this book, and not necessarily capture the full extent of the work in each site.

A.4.1 Cross-sectional Household Surveys

Household surveys were among the first study tools we administered in each of our sites to understand the types and distribution of water risks across households. The questionnaire was designed to gather baseline data on sociodemographic profile, asset ownership, housing structures, access to water and sanitation facilities, and development concerns/ priorities, with minor contextual modifications for each site. Between October 2017 and February 2018, we surveyed about 7,000 house-holds across our four sites – Polder 29 in Khulna district, along the Turag River and Tongi Khal in Dhaka and Gazipur districts, Mwingi-North subcounty in Kitui county, and Lodwar town and its peripheral areas in Turkana county. An addi-tional 2,000 households were surveyed in Polder 23 of Khulna district in January–February 2020.

   Household selection ensured uniform spatial representation either through stratified random sampling or through random transect walks. For example, in Dhaka, we covered all settlements within 1 km buffer zone along the river, using a random walk method to select every 10th household within each settlement. In Khulna, where mouza level (Tier 5 administrative boundary) census data was available, a stratified random sampling method was used to select 10 per cent of households from high-risk mouzas and 5 per cent from low-risk ones on Polder 29. The surveys were conducted in the local languages on ONA data collection platform via locally recruited enumerators, trained and supervised by research-ers at our local partner institutions. Data from these surveys are illustrated in Figures 3.4 and 5.3, showing the main sources of drinking water in Polder 29 and Lodwar town, respectively.

   The survey datasets provided the sampling frame for the water diaries in Khulna, Kitui, and Lodwar, with data on 'wealth status' (poor or non-poor) and 'concerns for water' ('water is costly', 'water is unsafe to drink' and 'others')

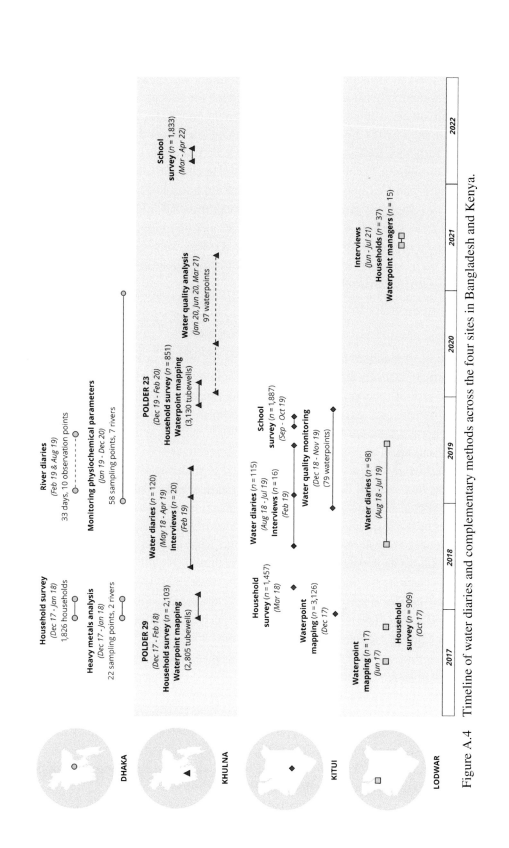

Figure A.4 Timeline of water diaries and complementary methods across the four sites in Bangladesh and Kenya.

being used as selection criteria. Wealth status was generated through principal component analysis of selected variables on education, housing structures, durable assets, and possession of consumable goods (for Kenya only). The factor loadings of the first principal component served as the weights of selected asset variables, while and the factor scores were regarded as the wealth index of each household. The median value was used to categorise households into 'poor' and 'non-poor' categories. During the survey, households also ranked their top three development priorities amongst 13 options that included healthcare, education, roads, water services, and flood protection amongst others. Those who expressed concerns about drinking water services were further asked to characterise their concerns in terms of quality (unsafe), quantity (insufficient), affordability (costly), and reliability.

A.4.2 Interviews

A sample of the diary households, 20 in Khulna and 15 in Kitui, were interviewed in February 2019 by the lead researcher (Sonia Hoque) to gain further insights into the drivers of recorded water choices. The quantitative diary data on household water choices and expenditures, for the number of weeks completed before the interviews, were used to select the households and guide the interview questions and discussions.

In Lodwar, the interviews were intended to understand how households navigate the dual risks of flash floods and unreliable or unavailable water services, and the challenges of relocating to peri-urban areas. Interviews involved in-depth narratives of the respondents' life histories, exploring their experiences of droughts, conflicts, or migration, their family dynamics, and income generating activities to better situate their water-related decisions within the broader struggles of daily life. Thirty-seven households from eight villages were interviewed in July and August 2021 by two local research assistants, who were recruited and trained remotely by the lead author. The villages were selected purposively to represent those affected by flash floods and those that the flood victims migrated to. While selecting households within each village, the research assistants ensured spatial representation, though in some instances, participant recruitment depended on availability during the time of visit. The interviews were conducted in the local language, with the research assistants taking written notes that they later translated and transcribed. In addition, a short, structured questionnaire was used to collect data on the household's sociodemographic profile at the end of each interview.

In Dhaka, interviews were conducted to understand the impacts of pollution exposure through monsoon flooding. Following exploratory visits, two sites were selected – a peri-urban community in Islampur along the Turag River and a rural community in Hazratpur, a few downstream of the Savar tannery estate along the Dhaleswari River. Ten interviews were conducted in each site by a research assistant in March 2023.

A.4.3 Water Infrastructure Mapping

One of the major drivers of source choice is the availability of waterpoints and their associated characteristics, such as distance, water quality, reliability, and costs. Water supply infrastructure was mapped in Khulna, Kitui, and Lodwar in multiple phases, recording the location, technical specifications, water usage patterns, and operation and maintenance information.

In Khulna, the water infrastructure audit comprised four components: (1) survey of all public and private tube wells in the southern half (high salinity area) of Polder 29 and Polder 23 ($n = 5,707$) (illustrated in Figure 3.2), and a selected sample in the northern half (low salinity area) of Polder 29; (2) a survey of all non-tube well sources including small piped schemes, ponds sand filters, rainwater reservoirs, community ponds and desalination plants in Polders 29 (illustrated in Figure 3.4) and Polder 23; (3) survey of all desalination plants in entire Paikgachha upazila (within which Polder 23 is located) and Dacope upazila; and (4) survey of all piped water schemes in Khulna district, excluding Khulna city corporation. The respondents were either owners or managers of these waterpoints, and provided data on the location, capital investments, source of funds, ownership, installation date, functionality, technical specifications, user payments, and management of the water infrastructure.

In Kitui, a total of 3,126 water supply infrastructure and unequipped sources were mapped over two phases, starting from Mwingi-North subcounty in August 2016, and extending to the entire Kitui county in November–December 2017. The combined survey covered 687 hand pumps, 460 piped schemes, 655 sand dams, 613 earth dams, 268 shallow wells, 89 rock catchments, 28 springs, and 52 incomplete boreholes, with a questionnaire similar to the one in Khulna though with a greater emphasis on breakdown and repair events.

Waterpoints in Lodwar were partially mapped by different researchers at different times and the data from these exercises were combined to illustrate the distribution of water supply infrastructure in the town (see Figure 5.6). The location of LOWASCO boreholes, piped network and kiosks were first mapped in June 2017 as part of Maxwell et al. (2020)'s MSc dissertation, followed by interviews with owners or managers of selected waterpoints in June 2021. The latter was intended to complement the in-depth household interviews in eight villages discussed earlier, covering seven functional LOWASCO kiosks, six water tanks owned by the county government or by private enterprises, and two handpumps.

For non-functional waterpoints, which included four handpumps, four kiosks, and three tanks, only the GPS coordinates and photos were recorded. This dataset was further updated during a field visit in February 2022 and separate infrastructure audit conducted in selected urban, peri-urban, and rural areas in Turkana Central subcounty in June 2023.

A.4.4 Tracking Waterpoint Expenditures and User Payments

Household water source choices have significant impacts on revenue collection for different water supply infrastructure and the ability of waterpoint managers to provide reliable operation and maintenance services. We trained owners or managers of 16 waterpoints in Khulna to keep records of all user payments and operation and maintenance expenses for a year from September 2019 to August 2020. These included four motorised tube wells used by water vendors, one piped water scheme, ten pond sand filters, and one reverse osmosis plant In Polder 29 (refer to results in Section 3.4). All managers were provided with logbooks and compensated for their time, with regular visits by field officers to monitor the data collection process.

A.4.5 River Water Quality Monitoring

A water quality monitoring system was developed and implemented by our colleagues at BUET to analyse the spatial and seasonal changes in the health of Greater Dhaka's River systems between 2017 and 2021. This involved two sub-studies – (1) monthly monitoring of physiochemical parameters, starting from the Turag River and Tongi Khal in 2017 and expanding to 58 sampling points covering the entire watershed in 2020–2021; (2) analysis of 18 heavy metals and biotoxicity at 22 sampling points in December 2017 and January 2018.

For the first sub-study, samples were collected from a boat at 2 m depth, followed by field analysis of temperature, pH, dissolved oxygen, oxidation-reduction potential, electrical conductivity, and total dissolved solids using portable meters. Laboratory analysis was conducted for colour, alkalinity, dissolved organic carbon, ammoniacal nitrogen, nitrate, phosphate, iron (ferrous), chloride, sulphate, and sulphide ions, and pathogens (total coliforms and *E. coli*). To present the relative changes in river health over monsoon, post-monsoon, and dry seasons in 2019 and 2020 (refer to Figure 2.3), a Water Quality Index was calculated for each of the river reaches using the method adopted by the Canadian Council of Ministers of the Environment. The index comprises 15 parameters, namely temperature, pH, electrical conductivity, dissolved oxygen, oxidation-reduction potential, turbidity, colour, alkalinity, iron, ammonia-nitrogen, nitrate, phosphate, sulphide, sulphate, and chloride.

For the second sub-study, concentrations of heavy metals were analysed at the Department of Earth Sciences, Oxford University through Inductively Coupled Plasma Mass Spectrometry. The heavy metal data was used by Oxford Molecular Biosensors to develop bacterially derived sensors to detect the bioavailability of specific chemicals and their overall cell damage toxicity (see Rampley et al., 2019).

A.4.6 Drinking Water Quality Analysis

Water samples were analysed for chemical and microbial contamination for water-points mapped in Khulna and Kitui. In Khulna, the electrical conductivity was measured in-situ for all functional tube wells using portable meters during the water infrastructure mapping in Polder 29 and Polder 23, covering 6,289 public and private tube wells (see Figure 3.2). Along with data on tube well depth, these salinity values were used to analyse the local hydrogeology, in terms of aquifer availability and groundwater quality (see Akhter et al., 2023). In Polder 29, a sample of 97 waterpoints, comprising deep tube wells, shallow tube wells and pond sand filters, were selected to understand seasonal variations in water quality across different infrastructure types. Water samples were collected by trained staff from icddr,b in January 2020, June 2020 and March 2021 and tested for arsenic, manganese, chlorides and *E. coli*, as shown in Figure 3.7.

In Mwingi-North subcounty, a water quality monitoring study was designed to assess the perceptions of drinking water safety and responses of waterpoint managers to detection of chemical or microbial contamination (refer to Nowicki et al., 2022). Between December 2019 and November 2020, water samples were collected monthly from 79 water sources, including 12 handpumps, 52 piped groundwater taps, from 25 schemes (including 10 mixed tanks with rainwater collection), 3 earth dams, 5 open wells, and 7 piped surface water taps from 4 schemes. These were analysed for chemical (pH, conductivity, turbidity, fluoride) and microbial (*E. coli* and total coliforms), of which the results for conductivity and *E. coli* are displayed in Figure 4.4 for selected waterpoints.

A.5 Water Services in Rural Schools in Khulna and Kitui

A total of 1,887 primary and secondary schools in Kitui County were surveyed in September–October 2019 (Hope et al., 2021b), followed by a similar survey of 1843 government, private and religious schools in Khulna district in March–April 2022. The surveys were aimed to understand the state of water services in schools, with details on technical specifications, repair and maintenance activities, water safety, and funding and expenditures for all water supply sources used. Standard questions on sanitation and hygiene by the Joint Monitoring Programme

(WHO/UNICEF, 2018) were also included to obtain a comprehensive baseline of water, sanitation and hygiene facilities. In Bangladesh, the survey also involved sanitary inspection of drinking water infrastructure. Data from the school survey in Bangladesh informed the projection of costs and logistics for scaling up SafePani from the pilot unions to the entire district. The Kitui survey involved 26 enumerators, trained and supervised by a locally based research manager. The survey in Khulna was conducted by a team of 70 enumerators, coordinated by field managers based at the REACH Polder 29 office and remotely supervised by the author (Sonia Hoque). Both surveys were conducted in ONA data collection platform using tablets and mobile phones.

References

Abdullahi, A. (1997). *Colonial policies and the failure of Somali Secessionism in the Northern Frontier District of Kenya Colony, c. 1890–1968*. Thesis submitted for the Degree of Master of Arts. South Africa: Rhodes University.

Aggarwal, P. K., K. Froehlich, A. Basu, R. Poreda, K. Kulkarni, S. Tarafdar, M. Ali, N. Ahmed, A. Hussain, and M. Rahman (2000). *A report on isotope hydrology of groundwater in Bangladesh: Implications for characterization and mitigation of arsenic in groundwater*. TC Project (BGD/8/016). Vienna: International Atomic Energy Agency (IAEA).

Ahmed, K. M. (2011). Groundwater contamination in Bangladesh. In: Grafton, R. Q. and K. Hussey (eds.) *Water resources planning and management*. Cambridge: Cambridge University Press.

Ahmed, S. U. (1986). *Dacca: A study in urban history and development*. London: Curzon Press.

Akall, G. (2021). Effects of development interventions on pastoral livelihoods in Turkana County, Kenya. *Pastoralism*, 11 (1), 23. https://doi.org/10.1186/s13570-021-00197-2

Akhter, T., M. Naz, M. Salehin, S. T. Arif, S. F. Hoque, R. Hope, and M. R. Rahman (2023). Hydrogeologic constraints for drinking water security in southwest coastal Bangladesh: Implications for Sustainable Development Goal 6.1. *Water*, 15 (13), 2333. https://doi.org/10.3390/w15132333

Ali, T. (2016, November 6). Time to declare Turag dead. *The Daily Star*, front page.

Armstrong, A., E. Dyer, J. Koehler, and R. Hope (2022). Intra-seasonal rainfall and piped water revenue variability in rural Africa. *Global Environmental Change*, 76, 102592. https://doi.org/10.1016/j.gloenvcha.2022.102592

Armstrong, A., R. Hope, and C. Munday (2021). Monitoring socio-climatic interactions to prioritise drinking water interventions in rural Africa. *npj Clean Water*, 4 (1), 10. https://doi.org/10.1038/s41545-021-00102-9

Baffoe, G., and S. Roy (2023). Colonial legacies and contemporary urban planning practices in Dhaka, Bangladesh. *Planning Perspectives*, 38 (1), 173–196. https://doi.org/10.1080/02665433.2022.2041468

Bakker, K. (2011). Privatizing water: Governance failure and the world's urban water crisis. In: *Privatizing Water*. Ithaca, NY:

Bartlett, R., and C. Milligan (2015). What is diary method? In: Crow, G. (ed.) *What is?* London: Bloomsbury Academic.

BBS (2023). Population and housing census 2022. Dhaka: Bangladesh Bureau of Statistics, Ministry of Planning.

BBS/UNICEF (2021). *Bangladesh MICS 2019: Water quality thematic report*. Bangladesh Bureau of Statistics, Government of Bangladesh and UNICEF.

BELA v. GoB (2003). *BELA v. Government of Bangladesh and others (Tannery Case)*. Supreme Court of Bangladesh, High Court Division.

Berg, A., H. Chhaparia, S. Hedrich, and K.-H. Magnus (2021, March 25). *What's next for Bangladesh's garment industry, after a decade of growth?* McKinsey & Company.

Bishop, S. (2015). Using water diaries to conceptualize water use in Lusaka, Zambia. *ACME: An International E-Journal for Critical Geographies*, 14 (3), 688–699.

Black, M. (1990). *From handpumps to health: The evolution of water and sanitation programmes in Bangladesh, India and Nigeria*. New York: United Nations Children's Fund.

Bolger, N., A. Davis, and E. Rafaeli (2003). Diary methods: Capturing life as it is lived. *Annual Review of Psychology*, 54 (1), 579–616.

Bourdieu, P. (1990). *The logic of practice*. Redwood City, CA: Stanford University Press.

Bowrey, T. (1905). *A geographical account of countries round the Bay of Bengal, 1669 to 1679*. Cambridge: The Hakluyt Society.

Broch-Due, V., and T. Sanders (1999). Rich man, poor man, administrator, beast: The politics of impoverishment in Turkana, Kenya, 1890–1990. *Nomadic Peoples*, 3 (2), 35–55. https://doi.org/10.3167/082279499782409389

Bukachi, S. A., D. O. Omia, M. M. Musyoka, F. M. Wambua, M. N. Peter, and M. Korzenevica (2021). Exploring water access in rural Kenya: Narratives of social capital, gender inequalities and household water security in Kitui county. *Water International*, 46 (5), 677–696. https://doi.org/10.1080/02508060.2021.1940715

Business Daily (2016, January 12). Property developers rush to invest in Turkana ahead of highway upgrade. Business Daily.

Bussi, G., S. Shawal, M. A. Hossain, P. G. Whitehead, and L. Jin (2023). Multibranch modelling of flow and water quality in the Dhaka river system, Bangladesh: Impacts of future development plans and climate change. *Water*, 15 (17), 3027. https://doi.org/10.3390/w15173027

BWDB-UNDP (1982). Groundwater survey: The hydrogeological conditions of Bangladesh. *UNDP Technical Report DP/UN/BGD-74-009/1*. Bangladesh Water Development Board (BWDB) and United Nations Development Programme (UNDP).

BWDB (2013). *Coastal embankment improvement project, phase I – Environmental impact assessment of Polder 32*. Dhaka: Bangladesh Water Development Board.

Byron, R. K., and M. Yousuf (2022). *World Bank to steer Dhaka rivers back to life. The Daily Star*.

Cates, M. E., M. H. Bishop, L. L. Davis, J. S. Lowe, and T. W. Woolley (2004). Clonazepam for treatment of sleep disturbances associated with combat-related posttraumatic stress disorder. *Annals of Pharmacotherapy*, 38 (9), 1395–1399.

Chintalapati, P., C. Nyaga, J. P. Walters, J. Koehler, A. Javernick-Will, R. Hope, and K. G. Linden (2022). Improving the reliability of water service delivery in rural Kenya through professionalized maintenance: A system dynamics perspective. *Environmental Science & Technology*, 56 (23), 17364–17374. https://doi.org/10.1021/acs.est.2c00939

Collins, D., J. Morduch, S. Rutherford, and O. Ruthven (2009). *Portfolios of the poor: How the world's poor live on $2 a day*. Oxford: Princeton University Press.

Dalton, P. R. (1976). A socioecological approach to the control of Schistosoma mansoni in St Lucia. *Bulletin of the World Health Organization*, 54 (5), 587.

Dalton, P. R., and D. Pole (1978). Water-contact patterns in relation to Schistosoma haematobium infection. *Bulletin of the World Health Organization*, 56 (3), 417.

Damery, S., G. Walker, J. Petts, and G. Smith (2008). Addressing environmental inequalities: water quality. *Science report SC020061/SR2*. Bristol, UK: Environment Agency.

Department of Economic and Social Affairs (2022). *World population prospects 2022: Summary of results*. New York: United Nations.

Derbyshire, S. F. (2020). *Remembering turkana: Material histories and contemporary livelihoods in north-western Kenya*. Oxford: Routledge.

DoE (2017). *Surface and groundwater quality report 2016*. Dhaka: Department of Environment, Ministry of Environment and Forest.

DPHE (2019). *Status of water points for the month of June 2019*. Dhaka: Department of Public Health and Engineering.

Dyer, E., and R. Washington (2021). Kenyan long rains: A subseasonal approach to process-based diagnostics. *Journal of Climate*, 34 (9), 3311–3326. https://doi.org/10.1175/JCLI-D-19-0914.1

Elliott, M., T. Foster, M. C. MacDonald, A. R. Harris, K. J. Schwab, and W. L. Hadwen (2019). Addressing how multiple household water sources and uses build water resilience and support sustainable development. *npj Clean Water*, 2 (6). https://doi.org/10.1038/s41545-019-0031-4

Ertsen, M., and K. Ngugi (2021). Ambivalent assets: The success of sand-storage dams for rainwater harvesting in Kitui County, Kenya. *Frontiers in Water*, 3, 676167. https://doi.org/10.3389/frwa.2021.676167

Etyang, H. (2019). Turkana residents lament water scarcity, blame county for laxity. *The Star*, 26 November 2019.

Etyang, H. (2021). Power disconnection plunges Lodwar into water shortage. *The Star*, 13 April 2021.

EurEau (2020). The governance of water services in Europe. The European Federation of National Associations of Water Services.

Fendorf, S., H. A. Michael, and A. van Geen (2010). Spatial and temporal variations of groundwater arsenic in South and Southeast Asia. *Science*, 328 (5982), 1123–1127.

Fischer, A. (2019). Constraining risk narratives: A multidecadal media analysis of drinking water insecurity in Bangladesh. *Annals of the American Association of Geographers*. https://doi.org/10.1080/24694452.2019.1570840

Fischer, A., R. Hope, A. Manandhar, S. Hoque, T. Foster, A. Hakim, M. S. Islam, and D. Bradley (2020). Risky responsibilities for rural drinking water institutions: The case of unregulated self-supply in Bangladesh. *Global Environmental Change*, 65, 102152. https://doi.org/10.1016/j.gloenvcha.2020.102152

Fischer, A., R. Hope, P. Thomson, S. F. Hoque, M. M. Alam, K. Charles, N. E. Achi, S. Nowicki, S. A. I. Hakim, M. S. Islam, M. Salehin, D. Bradley, M. Ibrahim, and M. E. H. Chowdhury (2021). Policy reform to deliver safely managed drinking water services for schools in rural Bangladesh. *REACH Working Paper 11*. Oxford: University of Oxford.

Foster, T. (2013). Predictors of sustainability for community-managed handpumps in sub-Saharan Africa: Evidence from Liberia, Sierra Leone, and Uganda. *Environmental Science & Technology*, 47 (21), 12037–12046. https://doi.org/10.1021/es402086n

Foster, T., R. Hope, C. Nyaga, J. Koehler, J. Katuva, P. Thomson, and N. Gladstone (2022). Investing in professionalized maintenance to increase social and economic returns from drinking water infrastructure in rural Kenya. *Policy Brief*. Sustainable WASH Systems Learning Program and REACH Programme.

Foster, T., and J. Willetts (2018). Multiple water source use in rural Vanuatu: Are households choosing the safest option for drinking? *International Journal of Environmental Health Research*, 28 (6), 579–589. https://doi.org/10.1080/09603123.2018.1491953

FSD Kenya (2014). *Kenya financial diaries Shilingi Kwa Shilingi – The financial lives of the poor*. Nairobi: Financial Sector Deepening (FSD) Kenya and The Gateway Financial Innovations for Savings (GAFIS).

GED (2015). Bangladesh Progress Report 2015. *Millennium Development Goals*. General Economics Division (GED), Bangladesh Planning Commission.

General Economics Division (2018). *Bangladesh Delta Plan 2100*. Dhaka: Bangladesh Planning Commission, Ministry of Planning.

Gorvett, Z. (2021). The ancient fabric that no one knows how to make. *BBC Future*, 17 March 2021.

Gramling, C. (2013). Kenyan find heralds new era in water prospecting. *Science*, 341 (6152), 1327–1327. https://doi.org/10.1126/science.341.6152.1327

Grasham, C. F., S. F. Hoque, M. Korzenevica, D. Fuente, K. Goyol, L. Verstraete, K. Mueze, M. Tsadik, G. Zeleke, and K. J. Charles (2022). Equitable urban water security: Beyond connections on premises. *Environmental Research: Infrastructure and Sustainability*, 2 (4), 045011. https://doi.org/10.1088/2634-4505/ac9c8d/meta

Grey, D., D. Garrick, D. Blackmore, J. Kelman, M. Muller, and C. Sadoff (2013). Water security in one blue planet: Twenty-first century policy challenges for science. *Philosophical Transactions of the Royal Society A: Mathematical, Physical and Engineering Sciences*, 371 (2002). https://doi.org/10.1098/rsta.2012.0406

Gunawansa, A., L. Bhullar, and S. F. Hoque (2013). Introduction. In: Gunawansa, A., and L. Bhullar (eds.) *Water governance: An evaluation of alternative architectures*. Cheltenham: Edward Elgar Publishing.

Halder, A. K., C. Tronchet, S. Akhter, A. Bhuiya, R. Johnston, and S. P. Luby (2010). Observed hand cleanliness and other measures of handwashing behavior in rural Bangladesh. *BMC Public Health*, 10 (1), 545. https://doi.org/10.1186/1471-2458-10-545

Halliday, S. (2001). *The great stink of London: Sir Joseph Bazalgette and the cleansing of the Victorian metropolis*, Cheltenham: The History Press.

Hamilton, K., B. Reyneke, M. Waso, T. Clements, T. Ndlovu, W. Khan, K. DiGiovanni, E. Rakestraw, F. Montalto, C. N. Haas, and W. Ahmed (2019). A global review of the microbiological quality and potential health risks associated with roof-harvested rainwater tanks. *npj Clean Water*, 2 (1), 7. https://doi.org/10.1038/s41545-019-0030-5

Haque, N. (2017). Exploratory analysis of fines for water pollution in Bangladesh. *Water Resources and Industry*, 18, 1–8. https://doi.org/10.1016/j.wri.2017.05.001

Harriden, K. (2013). Water diaries: Generate intra-household water use data–generate water use behaviour change. *Journal of Water Sanitation and Hygiene for Development*, 3 (1), 70–80. https://doi.org/10.2166/washdev.2013.015

Harvey, P., P. Ikumi, and D. Mutethia (2003). *Sustainable handpump projects in Africa*. Loughborough: Water Engineering and Development Centre, Loughborough University.

Harvey, P. A., and R. A. Reed (2007). Community-managed water supplies in Africa: sustainable or dispensable? *Community Development Journal*, 42 (3), 365–378. https://doi.org/10.1093/cdj/bsl001

Hirpa, F. A., E. Dyer, R. Hope, D. O. Olago, and S. J. Dadson (2018). Finding sustainable water futures in data-sparse regions under climate change: Insights from the Turkwel River basin, Kenya. *Journal of Hydrology: Regional Studies*, 19, 124–135. https://doi.org/10.1016/j.ejrh.2018.08.005

Hogg, R. (1982). Destitution and development: The Turkana of north west Kenya. *Disasters*, 6 (3), 164–168. https://doi.org/10.1111/j.1467-7717.1982.tb00531.x

Hong, S. C. (2018). Developing the Leather Industry in Bangladesh. *ADB Brief No. 102*. Asian Development Bank.

Hope, R., A. Fischer, S. F. Hoque, M. M. Alam, K. Charles, M. Ibrahim, E. H. Chowdhury, M. Salehin, Z. H. Mahmud, T. Akhter, P. Thomson, D. Johnson, S. A. Hakim, M. S. Islam, J. W. Hall, O. Roman, N. E. Achi, and D. Bradley (2021a). Policy reform for safe drinking water service delivery in Bangladesh. *REACH Working Paper 9*. Oxford: University of Oxford.

Hope, R., T. Foster, J. Koehler, and P. Thomson (2019). Rural water policy in Africa and Asia. In: Dadson, S. J., D. E. Garrick, E. C. P.-R. J. W. Hall, R. Hope, and J. Hughes (eds.) *Water science, policy, and management: A global challenge*. London: Wiley Blackwell.

Hope, R., J. Katuva, C. Nyaga, J. Koehler, K. Charles, S. Nowicki, E. Dyer, D. Olago, F. Tanui, A. Trevett, M. Thomas, and N. Gladstone (2021b). Delivering safely-managed water to schools in Kenya. *REACH Working Paper 8*. Oxford: University of Oxford.

Hope, R., and M. Rouse (2013). Risks and responses to universal drinking water security. *Philosophical Transactions of the Royal Society A: Mathematical, Physical and Engineering Sciences*, 371 (2002). https://doi.org/10.1098/rsta.2012.0417

Hoque, S. F. (2023). Socio-spatial and seasonal dynamics of small, private water service providers in Khulna district, Bangladesh. *International Journal of Water Resources Development*, 39 (1), 89–112. https://doi.org/10.1080/07900627.2021.1951179

Hoque, S. F., and R. Hope (2018). The water diary method – proof-of-concept and policy implications for monitoring water use behaviour in rural Kenya. *Water Policy*, 20 (4), 725–743. http://dx.doi.org/10.2166/wp.2018.179

Hoque, S. F., R. Hope, S. T. Arif, T. Akhter, M. Naz, and M. Salehin (2019). A social-ecological analysis of drinking water risks in coastal Bangladesh. *Science of The Total Environment*, 679, 23–34. https://doi.org/10.1016/j.scitotenv.2019.04.359

Hoque, S. F., and R. Hope (2020). Examining the economics of affordability through water diaries in coastal Bangladesh. *Water Economics and Policy*, 06 (03). https://dx.doi.org/10.1142/S2382624X19500115

Hoque, S. F., R. Peters, P. Whitehead, R. Hope, and M. A. Hossain (2021). River pollution and social inequalities in Dhaka, Bangladesh. *Environmental Research Communications*, 3 (9), 095003. https://doi.org/10.1088/2515-7620/ac2458

Hoque, S. F., and M. Shamsuddha (2024). Water Risks and Rural Development in Coastal Bangladesh. *The Oxford Encyclopedia of Water Resources Management and Policy*. https://doi.org/10.1093/acrefore/9780199389414.013.831

House of Commons (2022). Water quality in rivers. *Fourth Report of Session 2021–22*. Environmental Audit Committee.

Howard, G., J. Bartram, A. Williams, A. Overbo, D. Fuente, and J.-A. Geere (2020). *Domestic water quantity, service level and health*. Geneva: World Health Organization.

Islam, K. (2016). Our Story of Dhaka Muslin. *AramcoWorld*, p. 26.

Islam, M. S., and E. O'Donnell (2020). Legal rights for the Turag: Rivers as living entities in Bangladesh. *Asia Pacific Journal of Environmental Law*, 23 (2), 160–177.

Islam, S. S. (1985). The role of the state in the economic development of Bangladesh during the Mujib Regime (1972–1975). *The Journal of Developing Areas*, 19 (2), 185–208.

Khan, N. S., S. Shawal, M. A. Hossain, N. Tasnim, and Paul G. Whitehead (2024). Assessing flooding extent and potential exposure to river pollution from urbanizing peripheral rivers within Greater Dhaka watershed. *Unpublished manuscript*. Bangladesh University of Engineering and Technology (BUET).

Khan, S., Q. Cao, Y. Zheng, Y. Huang, and Y. Zhu (2008). Health risks of heavy metals in contaminated soils and food crops irrigated with wastewater in Beijing, China. *Environmental pollution*, 152 (3), 686–692.

Kisovi, L. M. (1992). Changing land use policy and population problems in Kitui district, Kenya. *Journal of Eastern African Research & Development*, 22, 92–104.

KNBS (2019a). *Kenya population and housing census 2019. Volume I: Population by county and subcounty*. Nairobi, Kenya: Kenya National Bureau of Statistics.

KNBS (2019b). *Kenya population and housing census 2019. Volume IV: Distribution of population by socio-economic characteristics*. Nairobi, Kenya: Kenya National Bureau of Statistics.

Koehler, J., C. Nyaga, R. Hope, P. Kiamba, N. Gladstone, M. Thomas, A. Mumma, and A. Trevett (2022). Water policy, politics, and practice: The case of Kitui County, Kenya. *Frontiers in Water*, 4. https://doi.org/10.3389/frwa.2022.1022730

Koehler, J., S. Rayner, J. Katuva, P. Thomson, and R. Hope (2018). A cultural theory of drinking water risks, values and institutional change. *Global Environmental Change*, 50, 268–277. https://doi.org/10.1016/j.gloenvcha.2018.03.006

Kookana, R. S., P. Drechsel, P. Jamwal, and J. Vanderzalm (2020). Urbanisation and emerging economies: Issues and potential solutions for water and food security. *Science of The Total Environment*, 732, 139057. https://doi.org/10.1016/j.scitotenv.2020.139057

Korzenevica, M., P. O. a. Ng'asike, M. Ngikadelio, D. Lokomwa, P. Ewoton, and E. Dyer (2024). From fast to slow risks: Shifting vulnerabilities of flood-related migration in Lodwar, Kenya. *Climate Risk Management*, 43, 100584. https://doi.org/10.1016/j.crm.2024.100584

Kumar, S., P. Lal, and A. Kumar (2021). Influence of super cyclone "Amphan" in the Indian subcontinent amid COVID-19 pandemic. *Remote Sensing in Earth Systems Sciences*, 4 (1), 96–103. https://doi.org/10.1007/s41976-021-00048-z

Lawson, C. C., G. K. LeMasters, and K. A. Wilson (2004). Changes in caffeine consumption as a signal of pregnancy. *Reproductive Toxicology*, 18 (5), 625–633.

Ligate, F., J. Ijumulana, A. Ahmad, V. Kimambo, R. Irunde, J. O. Mtamba, F. Mtalo, and P. Bhattacharya (2021). Groundwater resources in the East African Rift Valley: Understanding the geogenic contamination and water quality challenges in Tanzania. *Scientific African*, 13, e00831. https://doi.org/10.1016/j.sciaf.2021.e00831

Macharia, L. (2020). Turkwel Dam might overflow anytime from now – Water Resources Authority warns. *The Star*, 16 October 2020.

Masinde, K., M. Rouse, M. Jepkirui, and K. Cross (2021). *Guidance on Preparing Water Service Delivery Plans: A manual for small to medium-sized water utilities in Africa and similar settings*. London: International Water Association.

Maxwell, C., D. Olago, S. Dulo, and P. Odira (2020). Water Availability Analysis of Multiple Source Groundwater Supply Systems in Water Stressed Urban Centers: Case of Lodwar municipality, Kenya. *Journal of Civil & Environmental Engineering*, 10(2). https://doi.org/10.37421/mccr.2020.10.339

McCabe, J. T. (1990). Success and failure: The breakdown of traditional drought coping institutions among the Pastoral Turkana of Kenya. *Journal of Asian and African Studies*, 25 (3–4), 146–160. https://doi.org/10.1163/156852190X00021

McNicholl, D., and R. Hope (2024). Reducing uncertainty in corporate water impact: The role of Results-Based Contracting for drinking water supply. *Briefing note*. Oxford: University of Oxford and Uptime Global.

McNicholl, D., R. Hope, A. Money, A. Lane, A. Armstrong, M. Dupuis, A. Harvey, C. Nyaga, S. Womble, J. Allen, J. Katuva, T. Barbotte, L. Lambert, M. Staub, P. Thomson, and J. Koehler (2020). Results-Based Contracts for Rural Water Services. Uptime consortium.

Meth, P. (2003). Entries and omissions: Using solicited diaries in geographical research. *Area*, 35 (2), 195–205. http://dx.doi.org/10.1111/1475-4762.00263

Ministry of Jalshakti (2024). *Jal Jeevan Mission* [Online]. Department of Drinking Water & Sanitation, Ministry of Jalshakti, Government of India. Available: https://jaljeevanmission.gov.in/ (Accessed 25 February 2024).

Mirdha, R. U. (2023). Tannery CETP needs renovation even before offering full service. *The Daily Star*, 4 July 2023.

Mishra, R. R., and P. Upadhyay (2021). *Ganga: Re-imagining, rejuvenating, re-connecting*. New Delhi: Rupa Publications India.

Mkutu Agade, K. (2014). 'Ungoverned space' and the oil find in Turkana, Kenya. *The Round Table*, 103 (5), 497–515. https://doi.org/10.1080/00358533.2014.966497

MoEF (1997). Environmental Conservation Rules 1997. Dhaka, Bangladesh: Ministry of Environment and Forest (MoEF), Government of People's Republic of Bangladesh.

MoEFCC (2010). Bangladesh Environment Conservation (Amendment) Act, 2010. Dhaka, Bangladesh: Ministry of Environment and Forest (MoEF), Government of People's Republic of Bangladesh.

Mohammad, A. H., and M. Alauddin (2005). Trade liberalization in Bangladesh: The process and its impact on macro variables particularly export expansion. *The Journal of Developing Areas*, 39 (1), 127–150.

Mooi, E., and M. Sarstedt (2011). *A concise guide to market research: The process, data, and methods using IBM SPSS statistics*. Heidelberg: Springer.

Mottaleb, K. A., and T. Sonobe (2011). An inquiry into the rapid growth of the garment industry in Bangladesh. *Economic Development and Cultural Change*, 60 (1), 67–89. https://doi.org/10.1086/661218

Mumma, A. (2005). Kenya's new water law: an analysis of the implications for the rural poor. *International Workshop on African Water Laws: Plural Legislative Frameworks for Rural Water Management in Africa*. Johannesburg, South Africa.

Munday, C., N. Savage, R. G. Jones, and R. Washington (2023). Valley formation aridifies East Africa and elevates Congo Basin rainfall. *Nature*, 615, 276–279. https://doi.org/10.1038/s41586-022-05662-5

Munger, E. S. (1950). Water problems of Kitui District, Kenya. *Geographical Review*, 40 (4), 575–582. https://doi.org/10.2307/211103

Mwiria, K. (1990). Kenya's Harambee secondary school movement: The contradictions of public policy. *Comparative Education Review*, 34 (3), 350–368. https://doi.org/10.1086/446951

MWR (1999). Sessional Paper No. 1 of 1999 on National Policy on Water Resources Management and Development. Ministry of Water Resources, Republic of Kenya.

Ngware, M. W., E. N. Onsomu, and D. I. Muthaka (2007). Financing secondary education in Kenya: Cost reduction and financing options. *Education Policy Analysis Archives*, 15, 24–24. https://doi.org/10.14507/epaa.v15n24.2007

Nicholas, A. S. (2018). *Turkana-dassanech relations: Economic diversification and inter-communal conflicts, 1984–2015*. Thesis submitted for the degree of Master of Arts in Armed Conflict and Peace Studies, University of Nairobi.

Nowicki, S., S. A. Bukachi, S. F. Hoque, J. Katuva, M. M. Musyoka, M. M. Sammy, M. Mwaniki, D. O. Omia, F. Wambua, and K. J. Charles (2022). Fear, efficacy, and environmental health risk reporting: Complex responses to water quality test results in low-income communities. *International Journal of Environmental Research and Public Health*, 19 (1), 597.

Nyaga, C., R. Hope, K. Charles, S. Nowicki, J. Katuva, P. Mugo, S. Hoque, D. Olago, M. K. Thomas, and A. Mumma (2024). Guaranteeing safe drinking water services for public schools in Kenya: A costed professional service delivery model for Kitui County. *REACH Working Paper 14*. University of Oxford.

Nyaga, C. (2019). A Water Infrastructure Audit of Kitui County. Sustainable WASH Systems Learning Partnership, USAID.

Nyanchaga, E. N. (2016). *History of water supply and governance in Kenya (1895–2005) lessons and futures*. Tampere, Finland: Tampere University Press.

OAG (2019). Report of the Auditor-General on Lodwar Water and Sanitation Company Limited for the year ended 30 June, 2019. Office of the Auditor-General, Republic of Kenya.

OCHA (2023). Horn of Africa Drought Regional Humanitarian Overview & Call to Action (Revised 26 May 2023). UN Office for the Coordination of Humanitarian Affairs.

Olago, D., and F. Tanui (2023). Environmental Monitoring and Management Plan (EMMP) – Sustainable Use and Management of the Lodwar Alluvial Aquifer System, Turkana County, Kenya. REACH Kenya Programme (University of Nairobi and University of Oxford).

Parker, J. (2020). A wasted Eden: Colonial water management and ecological change in Kitui, Kenya 1948–63. *Les Cahiers d'Afrique de l'Est/The East African Review*, 55. https://doi.org/10.4000/eastafrica.1346

Paul, C. J., M. A. Jeuland, T. R. Godebo, and E. Weinthal (2018). Communities coping with risks: Household water choice and environmental health in the Ethiopian Rift Valley. *Environmental Science & Policy*, 86, 85–94. https://doi.org/10.1016/j.envsci.2018.05.003

Pauli, B. J. (2019). *Flint fights back: Environmental justice and democracy in the Flint water crisis.* Cambridge, Massachusetts: MIT Press. https://doi.org/10.7551/mitpress/11363.001.0001

Peters, R. (2022). *'Bringing Back Golden Bangladesh': Decentered Regulation and the Political Economy of Water Pollution.* Dissertation submitted for the Degree of Doctor of Philosophy, University of Oxford.

Price, H., E. Adams, and R. S. Quilliam (2019). The difference a day can make: The temporal dynamics of drinking water access and quality in urban slums. *Science of The Total Environment*, 671, 818–826. https://doi.org/10.1016/j.scitotenv.2019.03.355

Prüss, A. (1998). Review of epidemiological studies on health effects from exposure to recreational water. *International Journal of Epidemiology*, 27 (1), 1–9. https://doi.org/10.1093/ije/27.1.1

Quadir, F. (2000). The political economy of pro-market reforms in Bangladesh: Regime consolidation through economic liberalization? *Contemporary South Asia*, 9 (2), 197–212. https://doi.org/10.1080/713658731

Ram, P. K., A. K. Halder, S. P. Granger, T. Jones, P. Hall, D. Hitchcock, R. Wright, B. Nygren, M. S. Islam, J. W. Molyneaux, and S. P. Luby (2010). Is structured observation a valid technique to measure handwashing behavior? Use of acceleration sensors embedded in soap to assess reactivity to structured observation. *The American journal of tropical medicine and hygiene*, 83 (5), 1070–1076. https://doi.org/10.4269/ajtmh.2010.09-0763

Rampley, C. P. N., P. G. Whitehead, L. Softley, M. A. Hossain, L. Jin, J. David, S. Shawal, P. Das, I. P. Thompson, W. E. Huang, R. Peters, P. Holdship, R. Hope, and G. Alabaster (2019). River toxicity assessment using molecular biosensors: Heavy metal contamination in the Turag-Balu-Buriganga river systems, Dhaka, Bangladesh. *Science of The Total Environment*, 134760. https://doi.org/10.1016/j.scitotenv.2019.134760

Ray, I., and K. R. Smith (2021). Towards safe drinking water and clean cooking for all. *The Lancet Global Health*, 9 (3), e361–e365. https://doi.org/10.1016/S2214-109X(20)30476-9

REACH (2016). The FundiFix model: Maintaining rural water services. *REACH Working Paper*. University of Oxford.

REACH (2023a). Cost estimates for safe drinking water in schools and healthcare centres in Khulna District, Bangladesh. *Briefing Note*. University of Oxford.

REACH (2023b). The SafePani model: Delivering safe drinking water in schools and healthcare centres in Bangladesh. *Story of change*. University of Oxford.

REACH Dhaka (2023). *Water Quality of Rivers in Greater Dhaka* [Online]. Tableau Public Available: https://public.tableau.com/app/profile/reach.dhaka/viz/REACHDhaka/Dashboardmain?publish=yes.

Reimann, C., K. Bjorvatn, B. Frengstad, Z. Melaku, R. Tekle-Haimanot, and U. Siewers (2003). Drinking water quality in the Ethiopian section of the East African Rift Valley

I – data and health aspects. *Science of The Total Environment*, 311 (1), 65–80. https://doi.org/10.1016/S0048-9697(03)00137-2

Rocheleau, D. E., P. E. Steinberg, and P. A. Benjamin (1995). Environment, development, crisis, and crusade: Ukambani, Kenya, 1890–1990. *World Development*, 23 (6), 1037–1051. https://doi.org/10.1016/0305-750X(95)00016-6

Sagris, T., and J. Abbott (2015). *An analysis of industrial water use in Bangladesh with a focus on the textile and leather industries*. Washington, DC: 2030 Water Resources Group.

Schilling, J., T. Weinzierl, A. E. Lokwang, and F. Opiyo (2016). For better or worse: Major developments affecting resource and conflict dynamics in northwest Kenya. *Zeitschrift für Wirtschaftsgeographie*, 60 (1–2), 57–71. https://doi.org/10.1515/zfw-2016-0001

Shahid, S. (2011). Trends in extreme rainfall events of Bangladesh. *Theoretical and Applied Climatology*, 104 (3), 489–499. https://doi.org/10.1007/s00704-010-0363-y

Siddique, A., and S. Rahman (2019). Saving Buriganga a farce. *The Business Standard*, 11 December 2019.

Slootweg, R., M. Kooyman, P. de Koning, and M. van Schooten (1993). Water contact studies for the assessment of schistosomiasis infection risks in an irrigation scheme in Cameroon. *Irrigation and Drainage Systems*, 7 (2), 113–130. https://doi.org/10.1007/BF00880871

Sobhan, R. (1993). Structural maladjustment: Bangladesh's experience with market reforms. *Economic and Political Weekly*, 28 (19), 925–931.

Tanui, F., D. Olago, S. Dulo, G. Ouma, and Z. Kuria (2020). Hydrogeochemistry of a strategic alluvial aquifer system in a semi-arid setting and its implications for potable urban water supply: The Lodwar Alluvial Aquifer System (LAAS). *Groundwater for Sustainable Development*, 11, 100451. https://doi.org/10.1016/j.gsd.2020.100451

Tanui, F., D. Olago, G. Ouma, and Z. Kuria (2023). Hydrochemical and isotopic characteristics of the Lodwar Alluvial Aquifer System (LAAS) in Northwestern Kenya and implications for sustainable groundwater use in dryland urban areas. *Journal of African Earth Sciences*, 206, 105043. https://doi.org/10.1016/j.jafrearsci.2023.105043

Tavernier, J.-B., J. Phillips, H. Oldenburg, and E. Everard (1684). *Collections of travels through Turky into Persia, and the East-Indies*. London: Moses Pitt at the Angel in St. Pauls Church-yard.

The Daily Star (2017). BGMEA signs PaCT for green production. *The Daily Star*, 1 October 2017.

The Daily Star (2023). Bangladesh now home to half of top green factories worldwide. *The Daily Star*, 7 February 2023.

The Water Act (2002). Kenya Gazette Supplement No. 107 (Acts No. 9). Nairobi, Kenya: Republic of Kenya.

The Water Act (2016). Kenya Gazette Supplement No. 164 (Act No. 43). Nairobi, Kenya: Republic of Kenya.

Therkildsen, O. (1988). *Watering white elephants?: Lessons from donor funded planning and implementation of rural water supplies in Tanzania*. Uppsala: Scandinavian Institution of African Studies

Thomson, P., D. Bradley, A. Katilu, J. Katuva, M. Lanzoni, J. Koehler, and R. Hope (2019). Rainfall and groundwater use in rural Kenya. *Science of The Total Environment*, 649, 722–730. https://doi.org/10.1016/j.scitotenv.2018.08.330

Turbow, D. J., N. D. Osgood, and S. C. Jiang (2003). Evaluation of recreational health risk in coastal waters based on enterococcus densities and bathing patterns. *Environmental Health Perspectives*, 111 (4), 598–603. https://doi.org/10.1289/ehp.5563

Turkana County Water Act (2019). Kenya Gazette Supplement No. 7 (Turkana County Act No. 3). Nairobi: Republic of Kenya.

UN (2010). *Resolution adopted by the General Assembly on 28 July 2010 – The human right to safe drinking water and sanitation.* Geneva: Office of the United Nations High Commissioner for Human Rights (OHCHR).

UN (2015). *Report of the special rapporteur on the human right to safe drinking water and sanitation.* Geneva: Office of the United Nations High Commissioner for Human Rights (OHCHR).

UNDP (2006). *Human development report. Beyond scarcity: Power, poverty and the global water crisis.* New York: Palgrave Macmillan and United Nations Development Programme.

UNEP (2016). *A snapshot of the world's water quality – Towards a global assessment.* Nairobi: United Nations Environment Programme (UNEP).

UNEP (2021). Progress on ambient water quality. Tracking SDG 6 series: global indicator 6.3.2 updates and acceleration needs. Nairobi, Kenya: United Nations Environment Programme.

UNICEF/MICS (2019). Bangladesh Multiple Indicator Cluster Survey (MICS) – Round 6. UNICEF.

Villanueva, A. (2016). Urban River Use and Risks: A Study of Practice along the Turag River in Dhaka, Bangladesh. Dissertation submitted for the MSc degree in Water Science, Policy and Management, University of Oxford.

Wadira, S. O. (2020). *Hydrochemical Characteristics of Aquifers in Mwingi North-Kenya.* Dissertation submitted for the Master of Science in Geology (Hydrogeology and Groundwater Resources Management), University of Nairobi.

Wang, X., T. Sato, B. Xing, and S. Tao (2005). Health risks of heavy metals to the general public in Tianjin, China via consumption of vegetables and fish. *Science of the Total Environment*, 350 (1–3), 28–37. https://doi.org/10.1016/j.scitotenv.2004.09.044

WASREB (2019). *Guideline on provision of water and sanitation services for rural and underserved areas.* Nairobi: Water Services Regulatory Board (WASREB).

WASREB (2022). *IMPACT – A performance report of Kenya's water services sector – 2020/21.* Nairobi: Water Services Regulatory Board.

Whitehead, P., G. Bussi, M. A. Hossain, M. Dolk, P. Das, S. Comber, R. Peters, K. J. Charles, R. Hope, and S. Hossain (2018). Restoring water quality in the polluted Turag-Tongi-Balu river system, Dhaka: Modelling nutrient and total coliform intervention strategies. *Science of the Total Environment*, 631, 223–232. https://doi.org/10.1016/j.scitotenv.2018.03.038

Whitehead, P., G. Bussi, R. Peters, M. Hossain, L. Softley, S. Shawal, L. Jin, C. Rampley, P. Holdship, and R. Hope (2019). Modelling heavy metals in the Buriganga River System, Dhaka, Bangladesh: Impacts of tannery pollution control. *Science of the Total Environment*, 697, 134090. https://doi.org/10.1016/j.scitotenv.2019.134090

Whittington, D., J. Davis, L. Prokopy, K. Komives, R. Thorsten, H. Lukacs, A. Bakalian, and W. Wakeman (2009). How well is the demand-driven, community management model for rural water supply systems doing? Evidence from Bolivia, Peru and Ghana. *Water Policy*, 11 (6), 696–718. https://doi.org/10.2166/wp.2009.310

WHO/UNICEF (2017). *Progress on drinking water, sanitation and hygiene – 2017 update and SDG baseline.* Geneva: World Health Organization (WHO) and the United Nations Children's Fund (UNICEF) Joint Monitoring Programme (JMP).

WHO/UNICEF (2018). *Core questions and indicators for monitoring WASH in schools in the sustainable development goals.* Geneva: WHO/UNICEF Joint Monitoring Programme (JMP).

WHO/UNICEF (2020). The Measurement and Monitoring of Water Supply, Sanitation and Hygiene (WASH) Affordability. WHO-UNICEF Joint Monitoring Programme (JMP), the UN-Water Global Assessment and Analysis of Sanitation and Drinking-Water (GLAAS) and an Expert Group on WASH Affordability.

WHO/UNICEF (2023). *Progress on household drinking water, sanitation and hygiene 2000–2022: Special focus on gender*. New York: United Nations Children's Fund (UNICEF) and World Health Organization (WHO) Joint Monitoring Programme for Water Supply, Sanitation and Hygiene.

Wiseman, V., L. Conteh, and F. Matovu (2005). Using diaries to collect data in resource-poor settings: Questions on design and implementation. *Health Policy and Planning*, 20 (6), 394–404.

World Bank (2015). *Resettlement action plan for Marich Pass-Lodwar 196 km A1 road*. Washington, DC: World Bank.

World Bank (2018). *Benin – Rural water supply universal access program (English)*. Washington, DC: World Bank Group.

World Bank (2022). Bangladesh Environmental Sustainability and Transformation Project. Project Appraisal Document.

WSMTF (2023). *Water Services Maintenance Trust Fund – Impact summary, 2016–2023*. Nairobi, Kenya:

Wutich, A. (2006). *The effects of urban water scarcity on sociability and reciprocity in Cochabamba, Bolivia*. Doctor of Philosophy, University of Florida.

Wutich, A. (2009). Estimating household water use: A comparison of diary, prompted recall, and free recall methods. *Field Methods*, 21 (1), 49–68. https://doi.org/10.1177/1 525822X08325673

WWDR (2017). The United Nations World Water Development Report 2017. Wastewater: The Untapped Resource. Paris: UNESCO.

Index

Printed in the United States
by Baker & Taylor Publisher Services